Petroleum Engineer's Guide to Oil Field Chemicals and Fluids

Petroleum Engineer's Guide to Oil Field Chemicals and Fluids

Editor

Mr. Sanjay Patil

Petroleum Engineer's Guide to Oil Field Chemicals and Fluids

Edited by **Sanjay Patil**

Printed in 2017

ISBN: 978-1-68117-430-3

Library of Congress Control Number: 1-68117-430-8

© 2016 by
SCITUS Academics LLC,
616, Corporate Way, Suite 2, 4766,
Valley Cottage, NY 10989

www.scitusacademics.com

This book contains information obtained from highly regarded resources. Copyright for individual articles remains with the authors as indicated. All chapters are distributed under the terms of the Creative Commons Attribution License, which permits unrestricted use, distribution, and reproduction in any medium, provided the original author and source are credited.

Notice

Reasonable efforts have been made to publish reliable data and views articulated in the chapters are those of the individual contributors, and not necessarily those of the editors or publishers. Editors or publishers are not responsible for the accuracy of the information in the published chapters or consequences of their use. The publisher believes no responsibility for any damage or grievance to the persons or property arising out of the use of any materials, instructions, methods or thoughts in the book. The editors and the publisher have attempted to trace the copyright holders of all material reproduced in this publication and apologize to copyright holders if permission has not been obtained. If any copyright holder has not been acknowledged, please write to us so we may rectify.

Contents

Preface vii

Chapter 1 Theoretical Study of Hydrogen Bond Formation in Trimethylene Glycol-Water Complex 1
Snehanshu Pal and T. K. Kundu

Chapter 2 Applications of Nanofluids: Current and Future 39
Kaufui V. Wong and Omar De Leon

Chapter 3 Towards the Computational Design of Solid Catalysts 71
J. K. Nørskov, T. Bligaard, J. Rossmeisl, and C. H. Christensen

Chapter 4 Study and Application of Numerical Simulation of Deep Profile Control with Weak Gel 101
Zhou Yazhou and Yin Daiyin

Chapter 5 Environmental Impact Assessment of Oil and Gas Sector: a Case Study of Magurchara Gas Field 131
J. B. Alam, A. A. M. Ahmed, G. M. Munna, and A. A. M. Ahmed

Chapter 6 Ghana's Quest for Oil and Gas: Ecological Risks and Management Frameworks 145
Jian-Yi Liu, Jing Zhang, Yan-Li Liu, Xiao-Hua Tan, and Jie Zhang

Chapter 7 Hydrochemical and Hydrogeological Impact of Hydraulic Fracturing in the Karoo, South Africa 169
G. Steyl, and G. J. van Tonder

Chapter 8	**Microbial Hydrocarbon Degradation: Efforts to Understand Biodegradation in Petroleum Reservoirs**..................201
	Isabel Natalia Sierra-Garcia and Valéria Maia de Oliveira

Citations..................237

Index..................241

Preface

This manuscript is an extension and update of oil field chemicals, which appeared in 2003. The text focuses mainly on the organic chemistry of oil field chemicals. As indicated by the title, engineers with less background in organic chemistry are likely to use this text, so various sketches of the chemicals and additional explanations and comments are included in the text with which an educated organic chemist will certainly be familiar. The material presented here is a compilation from the literature, including patents, arranged in the order needed by a typical job. It starts with drilling fluids and related classes of compounds, such as fluid loss, bit lubricants, etc. Then it crosses over to the next major topics: cementing, fracturing, enhanced recovery, and it ends with ninelines and snill.

Editor

Chapter 1

Theoretical Study of Hydrogen Bond Formation in Trimethylene Glycol-Water Complex

Snehanshu Pal and T. K. Kundu

Department of Metallurgical and Materials Engineering, Indian Institute of Technology Kharagpur, West Bengal, Kharagpur 721302, India

ABSTRACT

A detailed quantum chemical calculation based study of hydrogen bond formation in trimethylene glycol- (TMG-) water complex has been performed by Hatree-Fock (HF) method, second-order Møller-Plesset perturbation theory (MP2), density functional theory (DFT), and density functional theory with dispersion function (DFT-D) using 6-31++G(d,p) basis set. B3LYP DFT-D, WB97XD, M06, and M06-2X functionals are used to capture highly dispersive hydrogen bond formation. Geometrical parameters, interaction energy, deviation of potential energy curve of hydrogen-bonded O–H from that of free O–H, natural bond orbital (NBO), atom in molecule (AIM), charge transfer, and red shift are investigated. It is observed that hydrogen bond between TMG and water molecule is stronger in case of TMG acting as proton donor compared to that of water acting as proton donor, and dilute TMG solution would inhibit water cluster formation.

INTRODUCTION

A hydrogen bond is an attractive donor-acceptor interaction, in which generally, the donor atoms are electronegative compared to hydrogen, and acceptor atoms have unshared lone pair electrons [1, 2]. The hydrogen bond has crucial impacts on many aspects of chemical and biological systems, and accordingly, hydrogen bond interaction has been an important research topic for several decades. The hydrogen bond also plays a significant role in formation of clathrate hydrate in marine sediments and below permafrost regions, which is considered to be significant future energy source [3, 4]. Global warming due to aleatory decomposition of methane hydrate [5] and hazards in petroleum industry owing to formation of gas hydrate in oil pipe line are of great concern [6, 7]. Controlled inhibition of gas hydrate formation is thus, very important, and various thermodynamic and kinetic inhibitors can break hydrogen-bonded network of clathrate structure by forming itself comparatively stronger hydrogen bond with water molecules of clathrate. Knowledge of hydrogen bond interaction is essential to identify potential gas hydrate inhibitor and design effective gas hydrate inhibitor. Trimethylene glycol being a polar compound can be a potential gas hydrate inhibitor as well as antifreeze reagent [7, 8].

Ab initio calculation is one of the most appropriate ways to obtain perspicuous understanding of hydrogen bond interaction and its impact on gas hydrate inhibition. Density functional theory (DFT) and atom in molecule (AIM) study of strong dihydrogen bonds [10] and resonance-assisted hydrogen bond [11] have been performed. Several theoretical studies of hydrogen bond interaction have been carried out for different systems like water complex [12,13], dichlorine monoxide-hydroxyl radical system [14], tetrahydrofuran-water complex [15,16], and methanol-water complex [17]. In one word, the literature of quantum chemical analysis of hydrogen bond interaction for various complex is well enriched [18–23]. Electronic structure-based studies on hydrogen bond formation between a molecule having two hydroxyl groups (like trimethylene glycol) and water have not been reported in the literature. Explicit study of interaction between trimethylene glycol and water is necessary to reveal the effect of a molecule having two hydroxyl groups on intermolecular and intramolecular hydrogen bond formation possibilities. This electronic structure-based insights on hydrogen bond formation can help in scientific understanding on application of trimethylene glycol as a gas hydrate inhibitor.

Our objective is to report a detailed theoretical analysis to comprehend the electronic nature of the hydrogen bond formation in trimethylene glycol-water system and its property using Hartree Fock, Møller-Plesset truncated at second-order (MP2), density function theory (DFT), and density functional theory with dispersion function (DFT-D). This study will help to conceptualize the nitty-gritty of hydrogen bond formation and its effect on vibrational spectra, natural bond orbital in trimethylene glycol-water complex.

COMPUTATIONAL DETAIL

Geometry optimization, determination of interaction energy, and natural bond orbital (NBO) analysis have been carried out using Hatree Fock (HF) [24] method, second-order Møller-Plesset perturbation theory (MP2) [25], density functional theory (DFT) [26, 27], and density functional theory with dispersion function (DFT-D) [28]. The calculations for DFT and DFT-D levels of theory have been performed using different functional, namely, B3LYP [29, 30], WB97XD [31], M06 [32], and M062X [32]. As polarity [33], of molecule has great influence on intermolecular hydrogen bonding, hydrogen bond-forming orbitals require larger space occupation [34]. Thus, diffuse and polarization functions augmented split valence 6-31++G(d,p) basis set is used for better description of molecular orbitals for geometry optimization and NBO analysis. Frequency calculation as well as AIM analysis have been performed using WB97XD/6-31++G(d,p) level of theory. Since hydrogen bonding is a kind of van der Waals type interaction, additional dispersion function with density functional theory, that is, DFT-D-based calculation has also been performed.

Interaction energy (ΔE_{HBF}) for hydrogen-bonded complex is calculated as the difference between the energy of hydrogen-bonded complex and the summation of the energies of each component monomer [35] as given in (1)

$$\Delta E_{HBF} = E_{complex} - \sum E_{component}, \quad (1)$$

where $E_{complex}$ and $E_{component}$ are optimized energy of hydrogen-bonded complex and each individual component monomer, respectively. Interaction energies have corrected for the basis set superposition error (BSSE) by virtue of counterpoise method [36]. A hydrogen-bonded complex is more stable if interaction energy is more negative compared to other hydrogen-bonded configurations.

Donor-acceptor interaction strength between filled orbital of the donor (Φ_i) and the empty orbital of acceptor (Φ_j) in case of natural bond orbital (NBO) analysis [37, 38] has been determined by second-order perturbation energy ($\Delta E_{ij}^{(2)}$) calculated using (2),

$$\Delta E_{ij}^{(2)} = 2 \frac{\langle \phi_i | F_{ij} | \phi_j \rangle^2}{\varepsilon_i - \varepsilon_j}, \quad (2)$$

where ε_i and ε_j are NBO energies, and F_{ij} is Fock matrix element between the i and j NBO orbitals. NBO analysis reveals the intra- and intermolecular interactions, and it is one of the appropriate methods for investigating hyperconjugative interactions [39].

Red shift in vibrational spectroscopy of conventional hydrogen-bonded structures arises from hyper-conjugation interaction [40]. Atom in molecule (AIM) study using Bader theory [41] has been performed as it is very effective for evaluating topological parameters of hydrogen bonds.

All the calculations have been carried out using Gaussian 09 software package [42]. Discovery Studio v3.1 of Accelrys software inc. is used for visualization of molecules. Vibrational frequency is calculated using 0.975 scaling factor [43].

RESULT AND DISCUSSION

Three possible conformations of trimethylene glycol denoted by TMG-1, TMG-2, and TMG-3 have been optimized using WB97XD/6-31++G(d,p) and shown in Figure 1. The TMG-2 conformation is found to be most stable as calculated relative energies of TMG-1 (3.77 kcal/mol) and TMG-3 (3.14 kcal/mol) isomers with respect to TMG-2 isomer are positive. This is because of intramolecular hydrogen bond formation in TMG-2 conformation. Detailed study of hydrogen bond interaction between water and TMG-2 conformation has been reported in this paper, and TMG-2 conformation is described as TMG. The optimized structures of TMG dimer, water dimer, TMG and one water complex considering either TMG or water as a proton donor (referred as TD and WD, resp.), and TMG + two water complex using WB97XD/6-31++G(d,p) calculation are shown in Figure 2. Here it is observed that intramolecular hydrogen bond distance (O12···H5) of TMG molecule increases in presence of water in WD conformation. It is also found that intermolecular hydrogen bond distances between TMG and water are less than intramolecular hydrogen bond distance (O12···H5) for TMG(TD)-one water complex and TMG + 2 water complexes. The intermolecular hydrogen bond in TMG + 1 water complex (TD) is smaller in length and consequently stronger than that of water dimer as evident in Figure 2. The calculated structural parameters using 6-31++G(d,p) basis set and different levels of theory are summarized in Tables 1(a) and 1(b). It is identified that B3LYP DFT-D [44], parameterized functional such as M06-, M06-2x-, and WB97XD-based methods which consider attractive dispersion force, show shorter hydrogen bond distances compared to HF theory-based calculation for all the systems. It is observed from Table 1(a), hydrogen bond angle values for intramolecular hydrogen bonds ($A_{O···H-O'}$, O12···H5–O4) of TMG molecule are less than hydrogen bond angle for intermolecular hydrogen bonds between TMG and water molecule in TMG + n water complex (n=1,2) for all calculation methods used in this paper. It is inferred based on hydrogen bond angle that the strength of intramolecular hydrogen bonds of TMG molecule are less compared to the strength of intermolecular hydrogen bonds between TMG and water molecule in TMG − water complex (n=1,2). The systems based on their dipole moment values in ascending order are water dimer < TMG < TMG + 1 water complex (TD) < TMG + 1 water complex (WD) < TMG + 2 water complex, for all the calculation procedures performed in this work, as evident in Table 1(a). Stronger intermolecular hydrogen bond formation enhances the dipole moment as hydrogen bond formation helps superposition of O···H moment and delocalization of πelectrons in hydrogen-bonded molecular complex [45].

Table 1: (a) Calculated hydrogen bond distances ($d_{O···H}$, Å), hydrogen bond angles ($A_{O···H}$, degree), dipole moment (D, debye) for single TMG, and TMG − n water complex (n = 1, 2) using 6-31++G(d,p) basis set and various methods. (b) Calculated hydrogen bond distances ($d_{O···H}$, Å), hydrogen bond angles ($A_{O···H}$, degree), dipole moment (D, debye) for water dimer, and TMG dimer using 6-31++G(d,p) basis set and various methods.

(a)

System	Parameters	MP2	WB97XD	MO6-2X	B3LYP DFTD	MO6	B3LYP	HF
TMG	$d_{O···H}$ O12–H5	2.01	2.02	2.02	2.03	2.04	2.04	2.12
	$A_{O···H}$ O12–H5–O4	137.17	137.36	135.62	138.00	136.10	137.45	133.52
	D	4.13	3.85	3.82	3.82	3.78	3.82	3.90
TMG + 1 water complex (TD)	$d_{O···H}$ O14–H13	1.89	1.87	1.88	1.85	1.90	1.89	2.01
	$d_{O···H}$ O12–H5	1.96	1.97	1.98	1.97	1.98	1.98	2.08
	$A_{O···H}$ O14–H13–O12	179.34	179.81	172.01	178.91	166.1	177.74	179.73
	$A_{O···H}$ O12–H5–O4	139.93	140.21	137.76	141.10	139.0	140.29	135.43
	D	5.45	5.14	5.61	4.96	5.77	5.04	5.35
TMG + 1 water complex (WD)	$d_{O···H}$ O12–H14	2.15	2.15	2.13	2.18	2.18	2.23	2.27
	$d_{O···H}$ O4–H16	2.29	2.18	2.16	2.09	2.15	2.21	2.46
	$d_{O···H}$ O12–H5	2.08	2.09	2.08	2.08	2.08	2.11	2.19
	$A_{O···H}$ O12–H14–O15	153.46	150.23	150.60	145.10	147.59	149.53	156.83
	$A_{O···H}$ O4–H16–O15	137.16	140.56	138.92	145.31	142.07	141.43	132.04
	$A_{O···H}$ O12–H5–O4	129.04	129.55	127.96	131.17	129.46	129.48	126.30
	D	5.84	5.56	5.44	5.58	5.51	5.67	5.55
TMG + 2 water complex	$d_{O···H}$ O12–H16	1.93	1.88	1.92	1.85	1.92	1.92	2.07
	$d_{O···H}$ O14–H18	2.03	1.98	2.00	1.96	2.01	2.01	2.16
	$d_{O···H}$ O4–H19	2.07	2.04	2.03	2.01	2.04	2.08	2.19
	$d_{O···H}$ O12–H5	2.08	2.09	2.05	2.10	2.06	2.13	2.20
	$A_{O···H}$ O12–H16–O14	167.66	171.38	159.54	170.48	160.10	168.10	170.42
	$A_{O···H}$ O14–H18–O17	159.62	160.19	161.95	160.01	161.85	160.78	158.30
	$A_{O···H}$ O4–H19–O17	153.14	152.70	151.64	154.50	153.11	154.97	154.52
	$A_{O···H}$ O12–H5–O4	134.18	134.34	133.78	134.72	134.33	133.69	129.97
	D	8.09	7.58	7.10	7.43	7.09	7.60	7.90

(b)

System	Parameters	MP2	WB97XD	MO6-2X	B3LYP DFTD	MO6	B3LYP	HF
Water dimer	$d_{O\cdots H}$, O1⋯H6	1.98	1.99	1.99	1.98	1.98	1.98	2.01
	$A_{O\cdots H}$, O1⋯H6–O2	175.61	175.04	172.63	174.40	175.05	174.21	176.21
	D	3.29	3.01	2.98	2.91	3.09	3.04	3.28
TMG dimer	$d_{O\cdots H}$, O12⋯H5	2.0	2.01	1.89	1.97	1.89	2.04	2.12
	$d_{O\cdots H}$, O25⋯H13	1.79	1.81	1.81	1.74	1.84	1.81	1.95
	$d_{O\cdots H}$, O4⋯H18	—	—	1.81	—	1.84	—	—
	$d_{O\cdots H}$, O12⋯H18	1.95	2.00	—	2.00	—	1.96	2.15
	$d_{O\cdots H}$, O17⋯H26	—	2.34	1.89	2.31	1.90	2.49	—
	$A_{O\cdots H}$, O12⋯H5–O4	136.76	136.49	144.21	140.60	144.81	139.91	133.99
	$A_{O\cdots H}$, O25⋯H13–O12	158.08	156.32	157.47	161.39	155.67	157.50	158.65
	$A_{O\cdots H}$, O4⋯H18–O17	—	—	157.46	—	155.76	—	—
	$A_{O\cdots H}$, O12⋯H18–O17	158.78	156.74	—	143.10	—	155.14	163.54
	$A_{O\cdots H}$, O17⋯H26–O25	—	114.80	144.19	116.28	144.77	108.57	—
	D	3.22	2.65	0.0011	1.31	0.0013	1.67	3.84

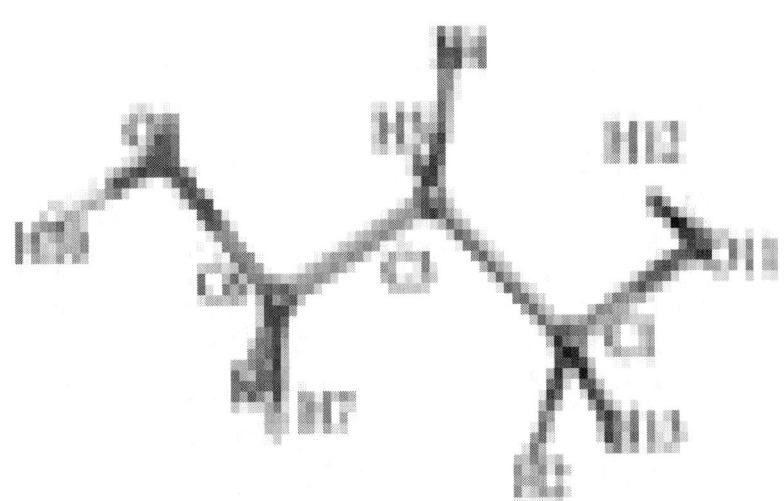

(a)

(b)

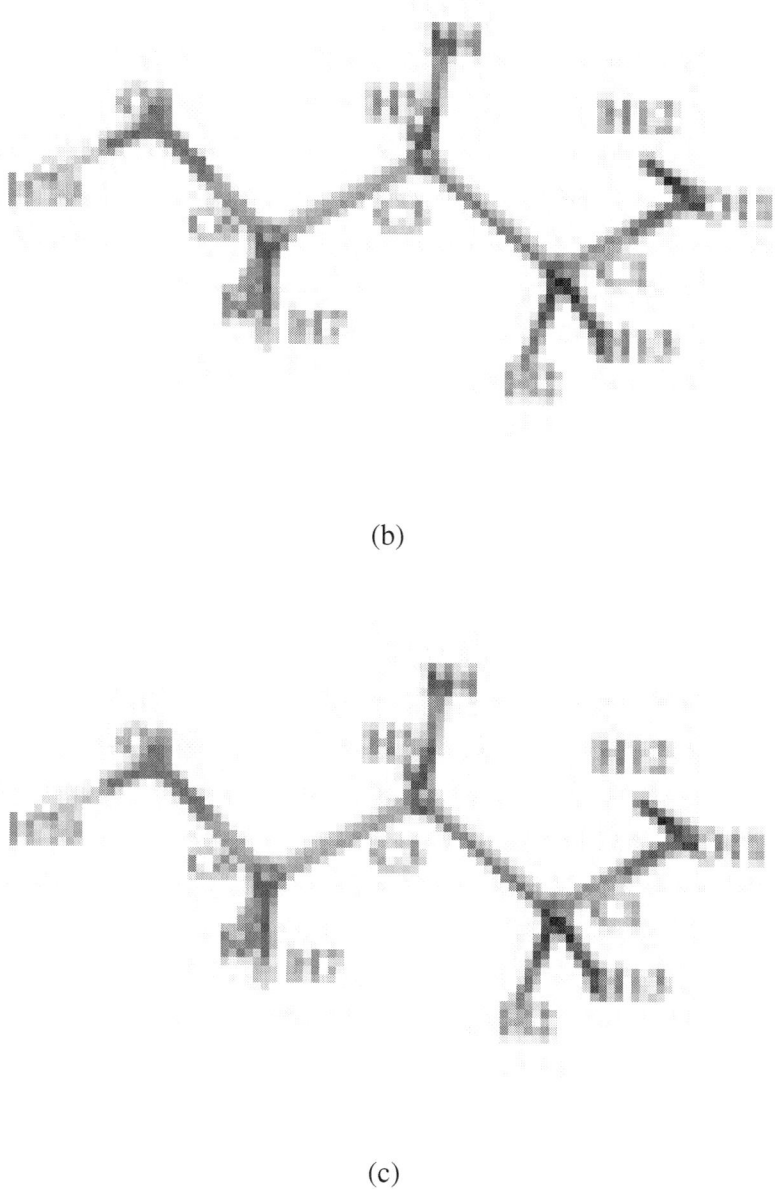

(c)

Figure 1: Optimized structures of different trimethylene glycol conformations such as (a) TMG-1, (b) TMG-2, and (c) TMG-3 using WB97XD/6-31++G(d,p) (colour legend: red = oxygen, black = carbon, and whitish grey = hydrogen, and black dotted line is hydrogen bond and hydrogen bond distance in Å).

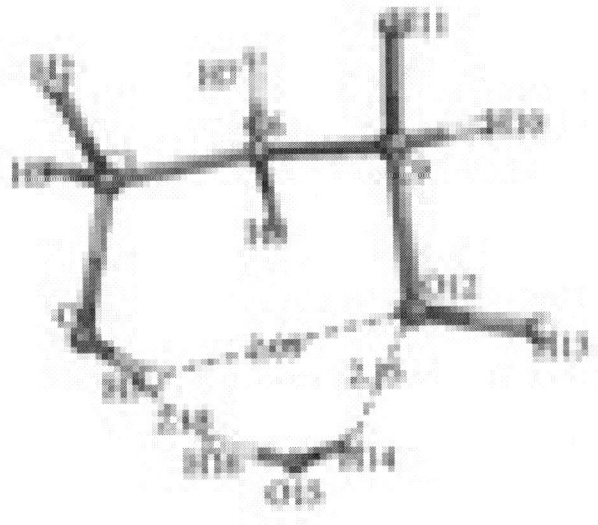

(a)

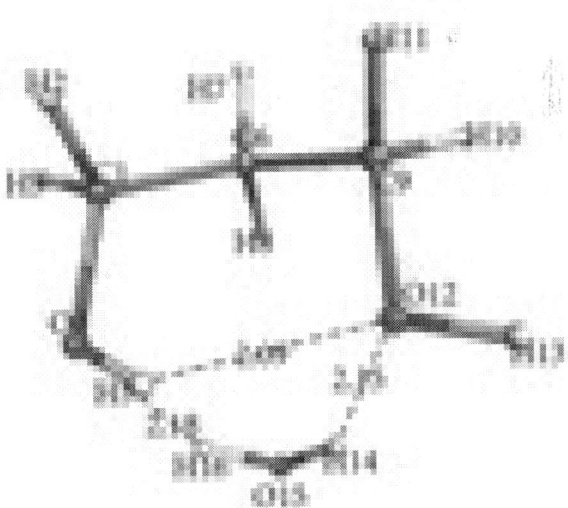

(b)

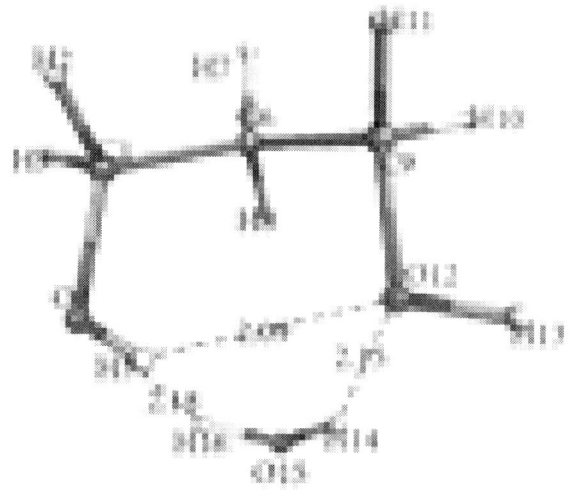

(c)

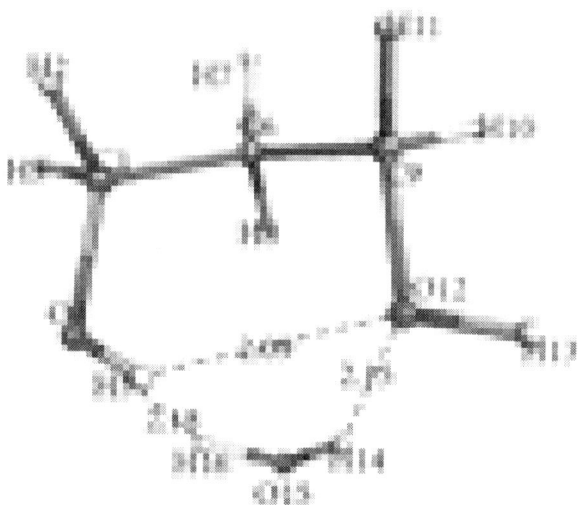

(d)

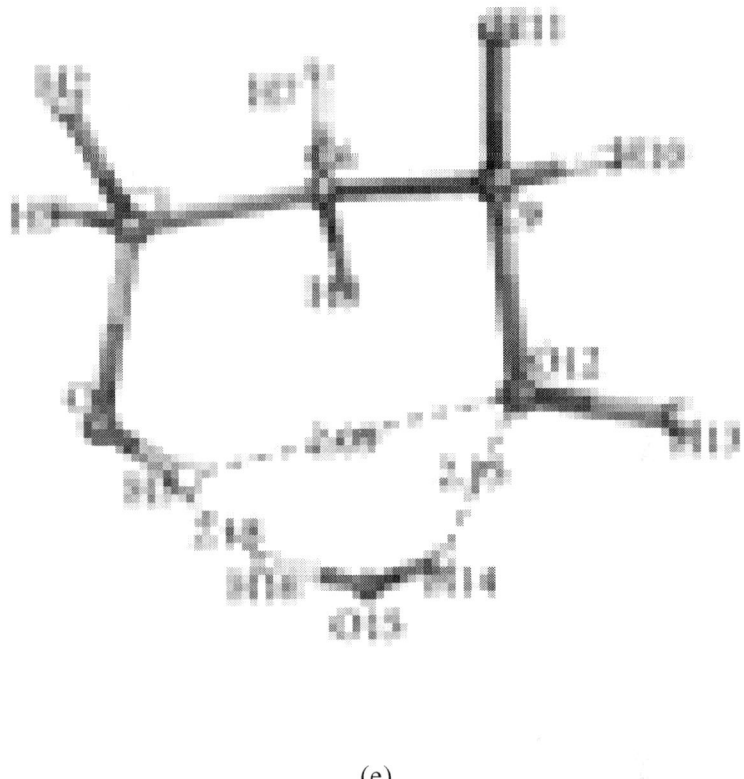

(e)

Figure 2: Optimized structures using WB97XD/6-31++G(d,p) of (a) trimethylene glycol (TMG) + 1 water complex (WD), (b) trimethylene glycol (TMG) + 1 water complex (TD), (c) trimethylene glycol (TMG) + 2 water complex, (d) water dimer, and (e) trimethylene glycol dimer (colour legend: red = oxygen, black = carbon, and whitish grey = hydrogen, and black dotted line is hydrogen bond and hydrogen bond distance in Å).

Calculated interaction energies with ($\Delta E_{HBF/CP}$) and without (ΔE_{HBF}) basis set superposition error (BSSE) correction (using counterpoise method) for TMG + n water complexes (n=1,2), TMG dimer, and water dimer along with number of hydrogen bonds formed are summarized in Table 2. The calculated interaction energies using HF method is less negative, indicating least hydrogen bond strength as it does not consider electron correlation. The hydrogen-bonded complex having more negative interaction energy should be more stable. Therefore, the ascending order with respect to stability is TMG + 1 water (WD) complex < TMG + 1 water (TD) complex < TMG + 2 water complex using HF, MP2, and B3LYP functional based methods, which exclude the dispersion term. The stability order in ascending sense using B3LYP DFT-D, WB97XD, M06, and M06-2X functional is water dimer < TMG + 1 water (TD) complex < TMG + 1 water (WD) complex < TMG dimer < TMG + 2 water complex. TMG + 1 water complex (TD) forms one intermolecular hydrogen bond (O14···H13), and TMG + 1 water (WD) complex form two intermolecular hydrogen bonds (O12···H14, O4···H16) as shown in Figures 2(b) and 2(c). Calculation methods using functionals having dispersion terms (B3LYP DFT-D, WB97XD, M06, and M06-2X) determine more negative interaction energy for a hydrogen-bonded complex compared to that obtained by HF, MP2 and B3LYP method as evident from Table 2. TMG has strong potential to form stable cluster with water molecules, and accordingly, dilute TMG solution would be useful as an inhibitor to restrict the formation of water cluster.

Table 2: Calculated interaction energy without correction (ΔE_{HBF}, kcal/mol), BSSE-corrected energy of hydrogen bond formation using counterpoise correction ($\Delta E_{HBF,CP}$, kcal/mol), hydrogen bond numbers for TMG + n water complex (n = 1, 2), TMG dimer, and water dimer using 6-31++G(d,p) basis set and various methods.

Systems	Calculation methods	ΔE_{HBF}	$\Delta E_{HBF,CP}$	No. of hydrogen bonds
	MP2	−7.65	−5.62	
	HF	−5.64	−4.94	
	B3LYP	−6.71	−5.77	
TMG + 1 water complex (TD)	B3LYP DFT-D	−7.79	−6.81	2
	WB97XD	−7.42	−6.51	
	M06	−7.00	−6.15	
	M062X	−7.59	−6.63	
	MP2	−7.46	−4.93	
	HF	−4.71	−3.88	

	B3LYP	−5.38	−4.58	
TMG + 1 water complex (WD)	B3LYP DFT-D	−8.92	−7.84	3
	WB97XD	−8.01	−7.01	
	M06	−7.39	−6.46	
	M062X	−8.48	−7.37	
	MP2	−16.39	−11.30	
	HF	−10.80	−9.24	
	B3LYP	−13.46	−11.44	
TMG + 2 water complex	B3LYP DFT-D	−18.32	−16.10	4
	WB97XD	−16.91	−14.81	
	M06	−16.33	−14.19	
	M062X	−17.90	−15.51	
	MP2	−9.46	−5.51	
	HF	−4.52	−3.51	
	B3LYP	−7.58	−6.61	
TMG dimer	B3LYP-DFTD	−13.38	−12.26	4
	WB97XD	−9.96	−8.90	
	M06	−14.69	−13.68	
	M062X	−15.18	−13.87	
	MP2	−6.39	−4.83	
	HF	−5.01	−4.36	
	B3LYP	−5.97	−5.19	
Water dimer	B3LYP-DFTD	−6.71	−5.95	1
	WB97XD	−6.35	−5.61	
	M06	−5.99	−5.25	
	M062X	−6.58	−5.80	

The potential energy curves for a free O–H (O12–H13) bond of single TMG molecule and hydrogen-bonded O–H (O12–H13) of TMG + 1 water complex (TD) are presented in Figures 3(a) and 3(b), respectively. The broadening of potential energy curve and appearance of asymmetrical double minimum in potential energy curve of hydrogen-bonded O–H reveal that a moderately strong hydrogen bond (O14···H13) is formed between TMG and water molecule [2]. The interaction energy barrier is high, which provides allowances for having various energetically lower protonic states [9].

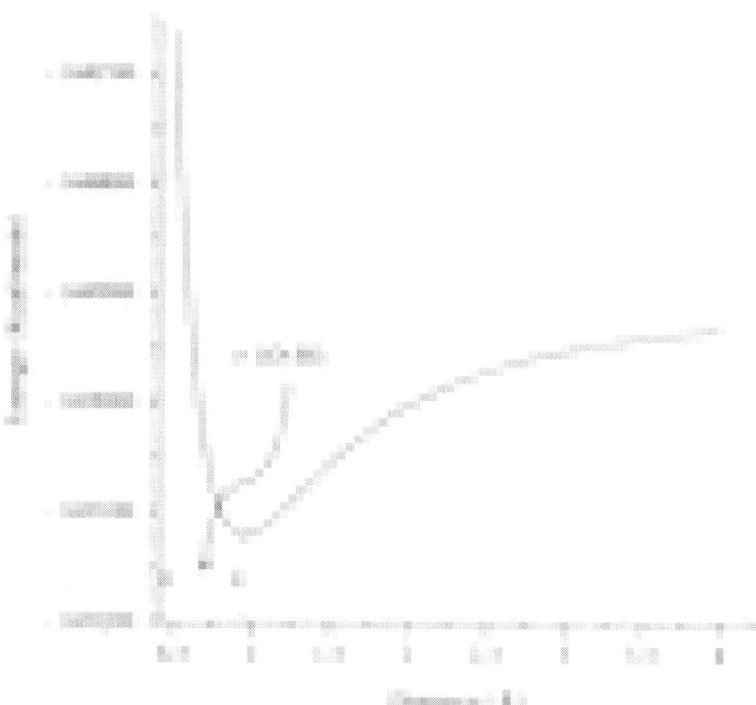

(a)

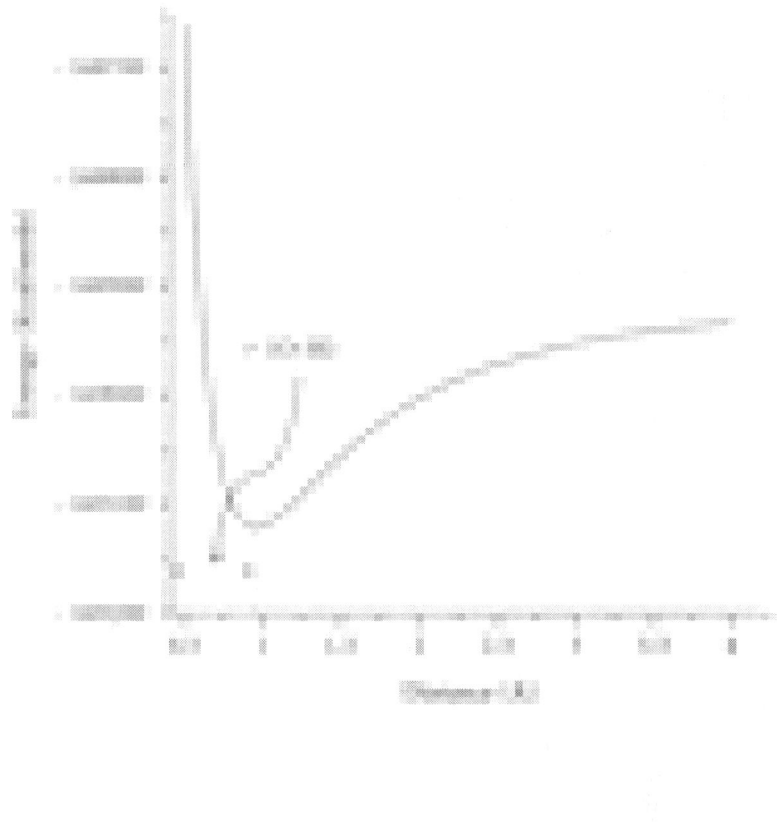

(b)

Figure 3: Calculated energy (kcal/mol) curve using WB97XD/6-31++G(d,p) for (a) a free bond of O–H (O12–H13) group of single TMG molecule (refer Figure 1(b)) and (b) hydrogen-bonded O–H (O12–H13) group of TMG + 1 water complex (TD) (refer Figure 2(b)).

Absolute Mullikan charge difference, absolute NBO charge difference, and absolute Chelpg charge difference between intermolecular hydrogen bond-forming oxygen and hydrogen atoms are obtained by taking absolute values of the difference between charge of oxygen and charge of hydrogen atoms and summarized in Table 3. As absolute charge differences between two atoms increase, the attractive electrostatic force between those two atoms also increases. Calculated absolute Mullikan charge difference, absolute NBO charge difference, and absolute Chelpg charge difference between intermolecular hydrogen bond-forming atoms (O14, H13) for TMG + 1 water complex (TD) are maximum and accordingly, forming strongest hydrogen bond compared to other systems for all the methods used in this paper, as shown in Table 3.

Table 3: Calculated absolute Mullikan charge difference, absolute NBO charge difference, and absolute Chelpg charge difference between hydrogen bond-forming atoms for TMG + n water complex (n = 1, 2) using 6-31++G(d,p) basis set and various methods

System	Methods	Hydrogen bond-forming atoms	Mullikan charge diff. (a.u.)	NBO charge diff. (a.u.)	Chelpg charge diff. (a.u.)
	MP2	O14 H13	1.25	1.54	1.39
	HF	O14 H13	1.23	1.53	1.43
	B3LYP	O14 H13	1.20	1.51	1.33
TMG + 1 water complex (TD)	B3LYP DFT-D	O14 H13	1.20	1.51	1.30
	WB97XD	O14 H13	1.20	1.52	1.33
	MO6	O14 H13	1.21	1.54	1.36
	MO62X	O14 H13	1.25	1.54	1.36
	MP2	O4 H16	1.02	1.34	1.14
		O12 H14	1.11	1.36	1.08
	HF	O4 H16	1.00	1.32	1.17
		O12 H14	1.09	1.34	1.11
	B3LYP	O4 H16	0.90	1.28	1.09
		O12 H14	1.01	1.30	1.03
TMG + 1 water complex (WD)	B3LYP DFT-D	O4 H16	0.91	1.28	1.09
		O12 H14	1.02	1.30	1.02
	WB97XD	O4 H16	0.93	1.29	1.10
		O12 H14	1.05	1.30	1.03

Theoretical Study of Hydrogen Bond Formation in Trimethylene...

	MO6	O4 H16	0.94	1.30	1.09
		O12 H14	1.06	1.32	1.03
	MO62X	O4 H16	0.97	1.30	1.09
		O12 H14	1.08	1.31	1.01
	MP2	O4 H19	1.06	1.34	1.17
		O12 H16	1.17	1.38	0.93
	HF	O4 H19	1.02	1.33	1.19
		O12 H16	1.14	1.36	0.97
	B3LYP	O4 H19	0.93	1.28	1.13
		O12 H16	1.06	1.32	0.85
TMG + 2 water complex	B3LYP DFT-D	O4 H19	0.94	1.28	1.13
		O12 H16	1.09	1.32	0.87
	WB97XD	O4 H19	0.96	1.29	1.13
		O12 H16	1.10	1.32	0.89
	MO6	O4 H19	0.99	1.31	1.00
		O12 H16	1.13	1.34	0.80
	MO62X	O4 H19	1.01	1.30	1.00
		O12 H16	1.15	1.34	0.79

Highest occupied molecular orbital (HOMO) and lowest unoccupied molecular orbital (LUMO) of TMG and water systems, simulated by WB97XD/6-31++G(d,p) method, are presented in Figure 4. The LUMO energies of TMG + n water complexes (n = 1, 2) are less compared to that of single TMG and water molecule. The LUMO of TMG + 1 water complex (TD) originates essentially from the LUMO of water with negligible contribution of antibonding orbital of TMG, but the HOMO of the same complex arises largely from the HOMO of TMG. On the other hand, for TMG + 1 water complex (WD) and TMG + 2 water complex, LUMO comes mainly from the LUMO of the TMG, and HOMO is from the intermixing of lone pairs of both TMG and water molecules. Intermolecular hydrogen bond (O14···H13) of TMG + 1 water complex (TD) has very high covalent character compared to two intermolecular hydrogen bonds (O12···H14, O4···H16) of TMG + 1 water complex (WD). It is also justified by the respective hydrogen bond lengths, that is, 1.87Å, 2.15Å, and 2.18Å for O14···H13, O12···H14, and O4···H16, respectively, as evident in Figures 2(b) and 2(c). In case of TMG + 2 water complex, the HOMO originates from major intermixing of lone pairs of TMG molecule and one water molecule (H19–O17–H18) and hardly any contribution from other water molecule (H16–O14–H15). It is also found from Figure 4, that the LUMO for TMG + 2 water complex originates from major intermixing of antibonding orbital of TMG molecule and one water molecule (H19–O17–H18). The covalent character is more prominent in one intermolecular hydrogen bond (O12···H16) compared to other hydrogen bond (O4···H19) in TMG + 2 water complex, and consequently the O12···H16 hydrogen bond is comparatively more strong, which is also supported by their hydrogen bond distances shown in Figure 1(d). Mixing of the HOMO of proton donor (O12–H13-bonding orbital of TMG) with the LUMO of proton acceptor (O14 of water molecule) in TMG + 1 water complex (TD) leads to decrease of electron density around O12–H13 bond.

(a)

(b)

Theoretical Study of Hydrogen Bond Formation in Trimethylene...

(c)

22 Petroleum Engineer's Guide to Oil Field Chemicals and Fluids

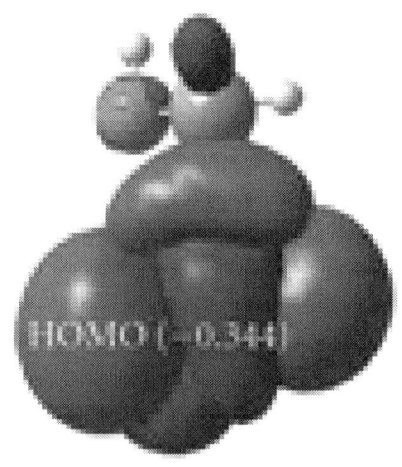

(d)

Theoretical Study of Hydrogen Bond Formation in Trimethylene... 23

(e)

(f)

Figure 4: Frontier orbitals (HOMO, LUMO energies is atomic unit) of (a) TMG monomer, (b) water monomer, (c) water dimer, (d) TMG dimer, (e) TMG + 1 water complex (TD), (f) TMG + 1 water complex (WD), and (g) TMG + 2 water complex by WB97XD/6-31++G(d,p) theory.

The calculated second-order perturbation energies and respective occupancies for selective donor-acceptor interactions relevant to hydrogen bond formation in single TMG molecule and TMG + 1 water (TD) complex from NBO analysis are given in Table 4. Calculated second-order perturbation energy of donor- (lone pair of O14) acceptor (antibonding orbital of O12–H13) interaction responsible for intermolecular hydrogen bonding is higher than that of donor- (lone pair of O12) acceptor (antibonding orbital of O4–H5) interaction responsible for intramolecular hydrogen bonding in TMG + 1 water complex (TD) according to all the methods used in this work. It is inferred that the intermolecular hydrogen bond is stronger than intra molecular hydrogen bond for TMG-water complex (TD) as supported by respective hydrogen bond distances.

Table 4: Calculated second-order perturbation energy ($\Delta E_{ij}^{(2)}$, kcal/mol) for TMG single and TMG − n water complex (n = 1, 2) using 6-31++G(d,p) basis set and various methods

System	Method	Donor	Occupancy of donor	Acceptor	Occupancy of acceptor	$\Delta E_{ij}^{(2)}$
	MP2	LP(2)O12	1.973	BD*(1) O4–H5	0.013	4.88 (Intra)
	HF	LP(1)O12	1.984	BD*(1) O4–H5	0.011	3.07 (Intra)
	B3LYP	LP(2)O12	1.979	BD*(1) O4–H5	0.019	4.72 (Intra)
TMG	B3LYP DFT-D	LP(2)O12	1.962	BD*(1) O4–H5	0.018	4.17 (Intra)
	WB97XD	LP(2)O12	1.963	BD*(1) O4–H5	0.018	5.09 (Intra)
	MO6	LP(2)O12	1.961	BD*(1) O4–H5	0.018	3.20 (Intra)
	MO62X	LP(2)O12	1.965	BD*(1) O4–H5	0.016	3.63 (Intra)

TMG + 1 water complex (TD)	MP2	LP(2)O12	1.968	BD*(1) O4–H5	0.017	7.28 (Intra)
		LP(2)O14	1.981	BD*(1) O12–H13	0.020	15.79 (Inter)
	HF	LP(1)O12	1.981	BD*(1) O4–H5	0.012	3.74 (Intra)
		LP(2)O14	1.988	BD*(1) O12–H13	0.014	10.12 (Inter)
	B3LYP	LP(2)O12	1.955	BD*(1) O4–H5	0.023	6.03 (Intra)
		LP(2)O4	1.978	BD*(1) O12–H13	0.030	14.30 (Inter)
	B3LYP DFT-D	LP(2)O12	1.952	BD*(1) O4–H5	0.025	7.07 (Intra)
		LP(2)O14	1.969	BD*(1) O12–H13	0.033	16.57 (Inter)
	WB97XD	LP(2)O12	1.956	BD*(1) O4–H	0.022	7.42 (Intra)
		LP(2)O14	1.972	BD*(1) O12–H13	0.030	18.04 (Inter)
	MO6	LP(2)O12	1.953	BD*(1) O4–H5	0.022	5.56 (Intra)
		LP(2)O14	1.976	BD*(1) O12–H13	0.026	13.09 (Inter)
	MO62X	LP(2)O12	1.960	BD*(1) O4–H5	0.019	5.05 (Intra)
		LP(2)O14	1.977	BD*(1) O12–H13	0.025	14.68 (Inter)

The charge transfer (CT) energies calculated using natural energy decomposition analysis (NEDA) for TMG, TMG + 1 water complex (TD), TMG + 1 water complex (WD), TMG + 2 water complex, water dimer, and TMG dimer are presented as bar chart in Figure 5. Charge transfer (CT) is a part of the stabilization energy of intermolecular interacting system [46], and it plays an important role in hydrogen bond formation [47, 48]. CT represents electron delocalization interaction between occupied molecular orbital of one molecule and unoccupied molecular orbital of another molecule. As TMG dimer shows lowest CT value compared to that of others, TMG dimer would be having the strongest intermolecular hydrogen bond interaction. Consequently,

TMG would be effective for inhibiting water cluster formation only when there is no favorable TMG dimer formation condition, that is, at low concentration of TMG.

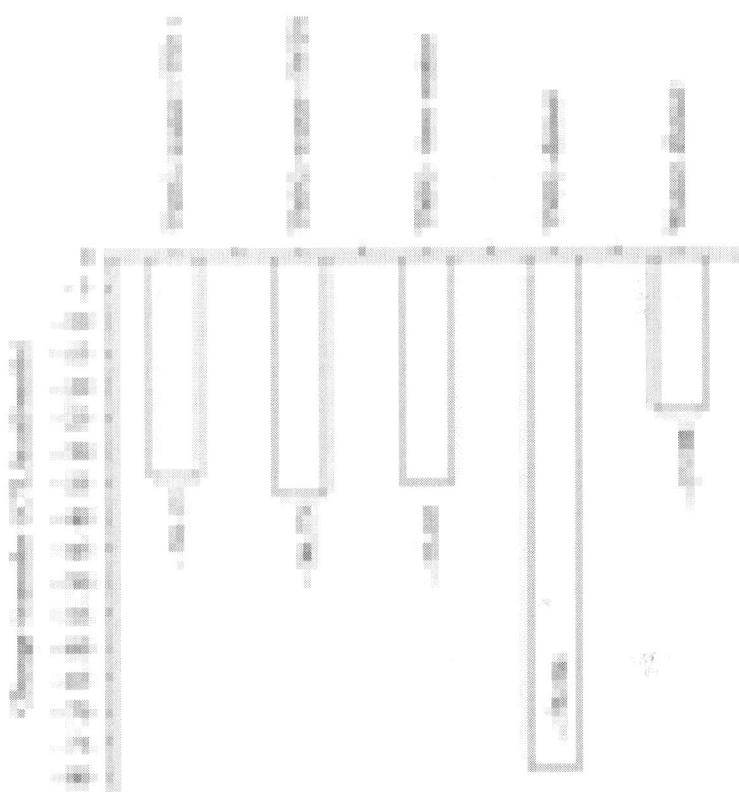

Figure 5: Bar chart of calculated charge transfer (CT, kcal/mol) by WB97XD/6-31++G(d,p) theory.

Calculated electron density contours with hydrogen bond critical point (HBCP), total electron density, and total Laplacian electron density at HBCP using AIM analysis for TMG molecule, water dimer, and TMG + 1 water complex (TD) using WB97XD/6-31++G(d,p) method are represented in Figure 6 and Table 5, respectively. One hydrogen bond critical point is determined for TMG molecule and water dimer, but two hydrogen bond critical points are found in TMG + 1 water complex (TD). The hydrogen bond critical point (HBCP) is a specific point between the donor and acceptor, where the gradient of electron density is zero, and it is essential evidence of hydrogen bond existence. TMG + 1 water complex (TD) has more covalence character and consequently more strength compared to water dimer as it has higher electron density at HBCP [41, 49] as evident in Table 5.

Table 5: Calculated total electron density ($\Sigma \rho(r_c)$, (e/a³)), total Laplacian electron density ($\Sigma \nabla^2 \rho(r_c)$, (e/a⁵)) for TMG molecule, water dimer, and TMG + 1 water (TD) complex at hydrogen bond critical point (HBCP) using WB97XD/6-31++G(d,p)

Systems	$\Sigma_{(rc)}$	$\Sigma_{2(rc)}$
Single TMG	0.0220	0.0701
Water dimer	0.0228	0.0645
TMG + 1 water complex (TD)	0.0291	0.0771

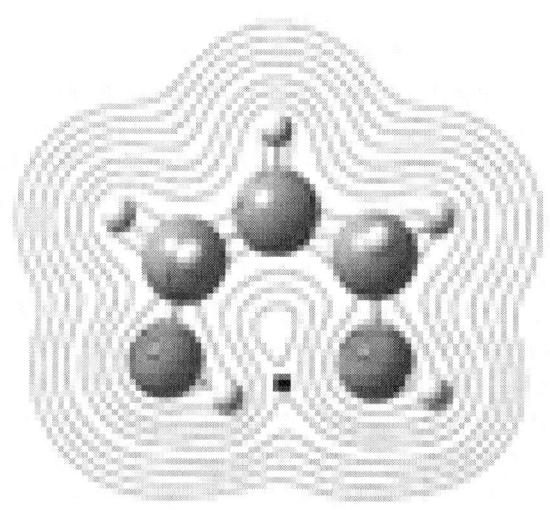

(a)

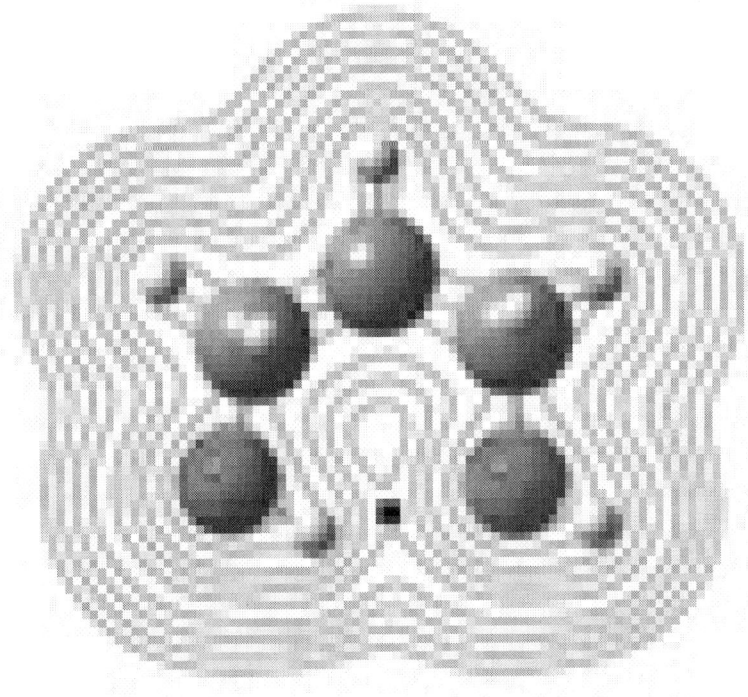

(b)

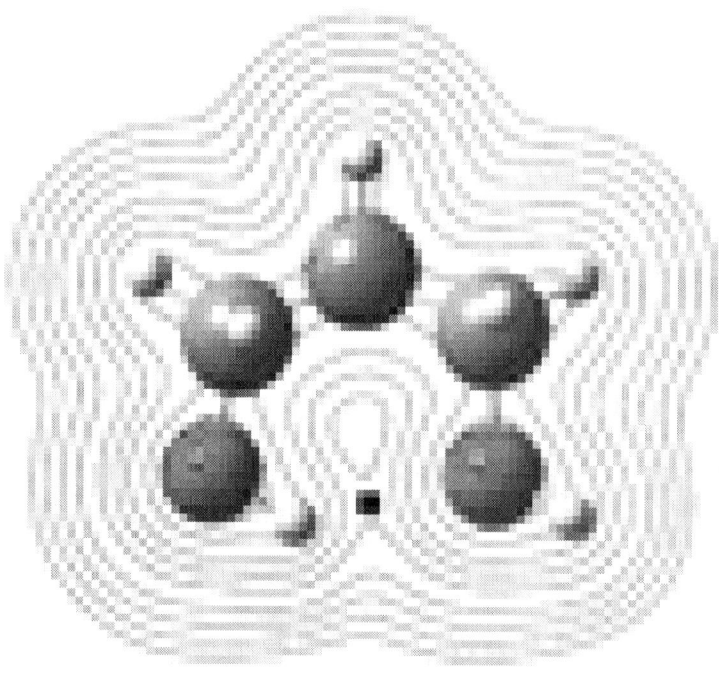

(c)

Figure 6: Contour map of the electron density for (a) single TMG molecule, (b) water dimer, (c) TMG + 1 water complex (TD) by WB97XD/6-31++G(d,p) theory. Hydrogen bond critical points are indicated by filled square symbol, ■ (colour legend: red = oxygen, black = carbon and whitish grey = hydrogen).

Calculated vibrational frequencies and IR intensities of O–H stretching of water and TMG molecule, TMG dimer, and TMG + n water complex system (n=1, 2) using WB97XD/6-31++G (d,p) are listed in Table 6. The red shift and intensity of hydrogen-bonded O–H stretching for TMG dimer is higher than that of TMG + n water complex (n=1, 2). It is also detected that the red shift and IR intensity of hydrogen-bonded O–H stretching of TMG molecule in TMG + 1 water (TD) system are higher than that of O–H stretching of water molecule in TMG + 1 water (WD) system. As higher values of red shift and intensity for hydrogen-bonded O–H bond stretching indicate stronger hydrogen bond, intermolecular hydrogen bond in TMG + 1 water (TD) system is stronger compared to the intermolecular hydrogen bond in TMG + 1 water (WD) system. Consequently, TMG has higher tendency to act as a proton donor to form hydrogen bond with water molecule.

Table 6: Calculated scaled vibrational frequency (cm^{-1}), red shift (cm^{-1}), IR intensity (km-mol^{-1}) of O–H bond stretching for water molecule, and TMG molecule and TMG + n water complex (n=1,2) using WB97XD/6-31++G(d,p) along with some experimental vibrational frequency (cm^{-1})

System	O–H bond stretching for water molecule				O–H bond stretching for TMG molecule		
	Scaled freq.	Red shift	IR intensity	Exp. vibrational frequency	Scaled freq.	Red shift	IR intensity
Water molecule	3802		8.2	3756 [9]			
TMG molecule					3757		144.43
TMG dimer					3519	238	585.61
TMG + 1 water complex (TD)	3797	5	13.87		3663	94	717.93
TMG + 1 water complex (WD)	3736	66	95.97		3754	3	80.39
TMG + 2 water complex	3665	137	512.89		3766	—	133.90

CONCLUSIONS

A thorough analysis of hydrogen bond formation in trimethylene glycol (TMG) + n water complex (n=1,2) has been performed based on calculated interaction energies, NBO, AIM, charge transfer, and red shift using HF, MP2, DFT, and DFT-D methods. TMG + 2 water complex, found to be most stable compared to TMG + 1 water complexes,

TMG dimer, and water dimer as per calculated interaction energies. For TMG + 1 water complex, stronger intermolecular hydrogen bond formed when TMG acts as a proton donor as per charge differences between respective hydrogen bond-forming atoms, NBO analysis, and red shifts of calculated vibrational spectra. The broadening as well as asymmetrical double minimum appearance in potential energy curve of hydrogen-bonded O–H reveals that a moderately strong hydrogen bond (O14···H13) is formed in TMG + 1 water complex (TD). Intermolecular hydrogen bond of TMG + 1 water complex (TD) has higher covalent character and accordingly, higher strength compared to that of TMG + 1 water complex (WD) as per HOMO-LUMO study. The hydrogen bond in TMG dimer is found to be stronger compared to other systems as per calculated charge transfer and red shift values. Very dilute TMG solution is recommended in order to break water cluster. This work illustrates electronic structure property correlation-based understandings of trimethylene glycol in aqueous solution and would help in designing inhibitors for water cluster/clathrate system like methane hydrate.

ACKNOWLEDGMENTS

This work is financially supported by Ministry of Earth Science, Government of India (Project no. MoES/16/48/09—RDEAS (MRDM5)). The authors also acknowledge Accelrys Inc. for providing free Discovery studio 3.1 visualization tool.

REFERENCES

1. P. Schuster and P. Wolschann, "Hydrogen bonding: from small clusters to biopolymers," Monatshefte fur Chemie, vol. 130, no. 8, pp. 947–960, 1999.
2. G. A. Jeffrey, An Introduction to Hydrogen Bonding, Oxford University Press, New York, NY, USA, 1997.
3. A. Demirbas, Methane Gas Hydrate, Springer, London, UK, 2010.
4. T. S. Collett, "Energy resource potential of natural gas hydrates," AAPG Bulletin, vol. 86, no. 11, pp. 1971–1992, 2002

5. P. Englezos, "Clathrate hydrates," Industrial and Engineering Chemistry Research, vol. 32, no. 7, pp. 1251–1274, 1993.
6. E. G. Hammerscht, "Formation of gas hydrates in natural gas transmission lines,"Industrial & Engineering Chemistry Research, vol. 26, no. 8, pp. 851–855, 1984.
7. J. K. Fink, Petroleum Engineer's Guide to Oil Field Chemicals and Fluids, Elsevier, Oxford, UK, 2012.
8. A. Wehner, R. Miller, G. Fenyvesi, J. W. DeSalvo, and M. Joerger, "Heat transfer compositions comprising renewable-based biodegradable 1, 3-propanediol," US patent 2007/0200088 A1, 2007.
9. V. May and O. Kühn, Charge and Energy Transfer Dynamics in Molecular Systems, Wiley-VCH, Weinheim, Germany, 2005.
10. S. J. Grabowski, T. L. Robinson, and J. Leszczynski, "Strong dihydrogen bonds—Ab initio and atoms in molecules study," Chemical Physics Letters, vol. 386, no. 1–3, pp. 44–48, 2004.
11. S. Wojtulewski and S. J. Grabowski, "DFT and AIM studies on two-ring resonance assisted hydrogen bonds," Journal of Molecular Structure, vol. 621, no. 3, pp. 285–291, 2003.
12. S. Pal and T. K. Kundu, "Dodecahedron methane hydrate cage structure—an Ab initio study," Journal of Petroleum Engineering and Technology, vol. 2, pp. 22–35, 2012.
13. D. Peeters, "Hydrogen bonds in small water clusters: a theoretical point of view," Journal of Molecular Liquids, vol. 67, pp. 49–61, 1995.
14. X. M. Zhou, Z. Y. Zhou, H. Fu, Y. Shi, and H. Zhang, "Density functional complete study of hydrogen bonding between the dichlorine monoxide and the hydroxyl radical ($Cl_2O \cdot HO$)," Journal of Molecular Structure, vol. 714, no. 1, pp. 7–12, 2005.
15. P. K. Sahu, A. Chaudhari, and S. L. Lee, "Theoretical investigation for the hydrogen bond interaction in THF-water complex," Chemical Physics Letters, vol. 386, no. 4–6, pp. 351–355, 2004.
16. P. K. Sahu and S. L. Lee, "Hydrogen-bond interaction in 1:1 complexes of tetrahydrofuran with water, hydrogen fluoride, and ammonia: a theoretical study,"Journal of Chemical Physics, vol. 123, no. 4, Article ID 044308, 9 pages, 2005.

17. A. Mandal, M. Prakash, R. M. Kumar, R. Parthasarathi, and V. Subramanian, "Ab Initio and DFT studies on methanol-water clusters," Journal of Physical Chemistry A, vol. 114, no. 6, pp. 2250–2258, 2010.
18. J. E. Del Bene, "An ab initio study of the structures and enthalpies of the hydrogen-bonded complexes of the acids H_2O, H_2S, HCN, and HCl with the anions OH-, SH-, CN-, and Cl-," Structural Chemistry, vol. 1, no. 1, pp. 19–27, 1990.
19. I. Alkorta, F. Blanco, P. M. Deyà et al., "Cooperativity in multiple unusual weak bonds,"Theoretical Chemistry Accounts, vol. 126, no. 1, pp. 1–14, 2010.
20. I. Mata, E. Molins, I. Alkorta, and E. Espinosa, "Topological properties of the electrostatic potential in weak and moderate N···H hydrogen bonds," Journal of Physical Chemistry A, vol. 111, no. 28, pp. 6425–6433, 2007.
21. J. B. Levy, N. H. Martin, I. Hargittai, and M. Hargittai, "Intra- and intermolecular hydrogen bonding in 2-phosphinylphenol: a quantum chemical study," Journal of Physical Chemistry A, vol. 102, no. 1, pp. 274–279, 1998.
22. O. V. Shishkin, I. S. Konovalova, L. Gorb, and J. Leszczynski, "Novel type of mixed O-H···N/O-H···π hydrogen bonds: monohydrate of pyridine," Structural Chemistry, vol. 20, no. 1, pp. 37–41, 2009.
23. V. Horváth, A. Kovács, and I. Hargittai, "Structural aspects of donor-acceptor interactions," Journal of Physical Chemistry A, vol. 107, no. 8, pp. 1197–1202, 2003.
24. C. C. J. Roothaan, "New developments in molecular orbital theory," Reviews of Modern Physics, vol. 23, no. 2, pp. 69–89, 1951.
25. M. Head-Gordon, J. A. Pople, and M. J. Frisch, "MP2 energy evaluation by direct methods," Chemical Physics Letters, vol. 153, no. 6, pp. 503–506, 1988.
26. P. Hohenberg and W. Kohn, "Inhomogeneous electron gas," Physical Review, vol. 136, no. 3, pp. B864–B871, 1964.
27. W. Kohn and L. J. Sham, "Self-consistent equations including exchange and correlation effects," Physical Review, vol. 140, no. 4, pp. A1133–A1138, 1965.

28. S. Grimme, "Accurate description of van der Waals complexes by density functional theory including empirical corrections," Journal of Computational Chemistry, vol. 25, no. 12, pp. 1463–1473, 2004.
29. A. D. Becke, "Density-functional exchange-energy approximation with correct asymptotic behavior," Physical Review A, vol. 38, no. 6, pp. 3098–3100, 1988.
30. C. Lee, W. Yang, and R. G. Parr, "Development of the Colle-Salvetti correlation-energy formula into a functional of the electron density," Physical Review B, vol. 37, no. 2, pp. 785–789, 1988. J. D. Chai and M. Head-Gordon, "Long-range corrected hybrid density functionals with damped atom-atom dispersion corrections," Physical Chemistry Chemical Physics, vol. 10, no. 44, pp. 6615–6620, 2008.
31. Y. Zhao and D. G. Truhlar, "The M06 suite of density functionals for main group thermochemistry, thermochemical kinetics, noncovalent interactions, excited states, and transition elements: two new functionals and systematic testing of four M06-class functionals and 12 other functionals," Theoretical Chemistry Accounts, vol. 120, no. 1–3, pp. 215–241, 2008.
32. P. C. Hariharan and J. A. Pople, "The influence of polarization functions on molecular orbital hydrogenation energies," Theoretica Chimica Acta, vol. 28, no. 3, pp. 213–222, 1973.
33. J. Chandrasekhar, J. G. Andrade, and P. Von Ragué Schleyer, "Efficient and accurate calculation of anion proton affinities," Journal of the American Chemical Society, vol. 103, no. 18, pp. 5609–5612, 1981.
34. M. S. Gordon and J. H. Jensen, "Understanding the hydrogen bond using quantum chemistry," Accounts of Chemical Research, vol. 29, no. 11, pp. 536–543, 1996.
35. S. F. Boys and F. Bernardi, "The calculation of small molecular interactions by the differences of separate total energies. Some procedures with reduced errors," Molecular Physics, vol. 19, no. 4, pp. 553–566, 1970.
36. F. Weinhold and C. R. Landis, "Natural bond orbitals and extensions of localized bonding concepts," Chemistry Education Research and Practice, vol. 2, pp. 91–104, 2001.

37. E. D. Gledening, A. E. Reed, J. A. Carpenter, and F. Weinhold, NBO. version 3.1.
38. A. E. Reed, L. A. Curtiss, and F. Weinhold, "Intermolecular interactions from a natural bond orbital, donor-acceptor viewpoint," Chemical Reviews, vol. 88, no. 6, pp. 899–926, 1988.
39. A. Y. Li, "Chemical origin of blue- and red shifted hydrogen bonds: intra-molecular hyper-conjugation and its coupling with intermolecular hyper-conjugation," Journal of Chemical Physics, vol. 126, pp. 154102–154111, 2007.
40. R. F. W. Bader, "Atoms in molecules," Accounts of Chemical Research, vol. 18, pp. 9–15, 1985.
41. M. J. Frisch, G. W. Trucks, H. B. Schlegel, et al., "Gaussian 09, Revision (B.01)," Gaussian Inc., Wallingford CT, 2010.
42. I. M. Alecu, J. Zheng, Y. Zhao, and D. G. Truhlar, "Computational thermochemistry: scale factor databases and scale factors for vibrational frequencies obtained from electronic model chemistries," Journal of Chemical Theory and Computation, vol. 6, no. 9, pp. 2872–2887, 2010.
43. B. Civalleri, C. M. Zicovich-Wilson, L. Valenzano, and P. Ugliengo, "B3LYP augmented with an empirical dispersion term (B3LYP-D*) as applied to molecular crystals,"CrystEngComm, vol. 10, no. 4, pp. 405–410, 2008.
44. A. E. Lutskii and N. I. Gorokhova, "Intramolecular hydrogen bonds and molecular dipole moments," Theoretical and Experimental Chemistry, vol. 4, no. 6, pp. 532–534, 1971.
45. H. Umeyama and K. Morokuma, "Origin of alkyl substituent effect in the proton affinity of amines, alcohols, and ethers," Journal of the American Chemical Society, vol. 98, no. 15, pp. 4400–4404, 1976.
46. H. Umeyama and K. Morokuma, "The origin of hydrogen bonding. An energy decomposition study," Journal of the American Chemical Society, vol. 99, no. 5, pp. 1316–1332, 1977.
47. A. Van der Vaart and K. M. Merz Jr., "Charge transfer in small hydrogen bonded clusters," Journal of Chemical Physics, vol. 116, no. 17, pp. 7380–7388, 2002.

48. S. J. Grabowski, Hydrogen Bonding-New Insights, Springer, Dordrecht, The Netherlands, 2006.

Chapter 2

Applications of Nanofluids: Current and Future

Kaufui V. Wong and Omar De Leon

Mechanical and Aerospace Engineering Department, University of Miami, Coral Gables, FL 33124, USA

ABSTRACT

Nanofluids are suspensions of nanoparticles in fluids that show significant enhancement of their properties at modest nanoparticle concentrations. Many of the publications on nanofluids are about understanding their behavior so that they can be utilized where straight heat transfer enhancement is paramount as in many industrial applications, nuclear reactors, transportation, electronics as well as biomedicine and food. Nanofluid as a smart fluid, where heat transfer can be reduced or enhanced at will, has also been reported. This paper focuses on presenting the broad range of current and future applications that involve nanofluids, emphasizing their improved heat transfer properties that are controllable and the specific characteristics that these nanofluids possess that make them suitable for such applications.

INTRODUCTION

Nanofluids are dilute liquid suspensions of nanoparticles with at least one of their principal dimensions smaller than 100 nm. From previous investigations, nanofluids have been found to possess enhanced thermophysical properties such as thermal conductivity, thermal diffusivity, viscosity and convective heat transfer coefficients compared to those of base fluids like oil or water [1–6]

From the current review, it can be seen that nanofluids clearly exhibit enhanced thermal conductivity, which goes up with increasing volumetric fraction of nanoparticles. The current review does concentrate on this relatively new class of fluids and not on colloids which are nanofluids because the latter have been used for a long time. Review of experimental studies clearly showed a lack of consistency in the reported results of different research groups regarding thermal properties [7, 8]. The effects of several important factors such as particle size and shapes, clustering of particles, temperature of the fluid, and dissociation of surfactant on the effective thermal conductivity of nanofluids have not been studied adequately. It is important to do more research so as to ascertain the effects of these factors on the thermal conductivity of wide range of nanofluids.

Classical models cannot be used to explain adequately the observed enhanced thermal conductivity of nanofluids. Recently most developed models only include one or two postulated mechanisms of nanofluids heat transfer. For instance, there has not been much fundamental work reported on the determination of the effective thermal diffusivity of nanofluids nor heat transfer coefficients for nanofluids in natural convection [9].

There is a growth is the use of colloids which are nanofluids in the biomedical industry for sensing and imaging purposes. This is directly related to the ability to design novel materials at the nanoscale level alongside recent innovations in analytical and imaging technologies for measuring and manipulating nanomaterials. This has led to the fast development of commercial applications which use a wide variety of manufactured nanoparticles. The production, use and disposal of manufactured nanoparticles will lead to discharges to air, soils and water systems. Negative effects are likely and quantification and minimization of these effects on environmental health is necessary.

True knowledge of concentration and physicochemical properties of manufactured nanoparticles under realistic conditions is important to predicting their fate, behavior and toxicity in the natural aquatic environment. The aquatic colloid and atmospheric ultrafine particle literature both offer evidence as to the likely behavior and impacts of manufactured nanoparticles [10], and there is no pretense that a review duplicating similar literature about the use of colloids which are also nanofluids is attempted in the current review.

Owing to their enhanced properties as thermal transfer fluids for instance, nanofluids can be used in a plethora of engineering applications ranging from use in the automotive industry to the medical arena to use in power plant cooling systems as well as computers.

HEAT TRANSFER APPLICATIONS

Industrial Cooling Applications

Routbort et al. [11] started a project in 2008 that employed nanofluids for industrial cooling that could result in great energy savings and resulting emissions reductions. For U.S. industry, the replacement of cooling and heating water with nanofluids has the potential to conserve 1 trillion Btu of energy. For the U.S. electric power industry, using nanofluids in closed-loop cooling cycles could save about 10–30 trillion Btu per year (equivalent to the annual energy consumption of about 50,000–150,000 households). The associated emissions reductions would be approximately 5.6 million metric tons of carbon dioxide; 8,600 metric tons of nitrogen oxides; and 21,000 metric tons of sulfur dioxide.

For Michelin North America tire plants, the productivity of numerous industrial processes is constrained by the lack of facility to cool the rubber efficiently as it is being processed. This requires the use of over 2 million gallons of heat transfer fluids for Michelin's North American plants. It is Michelin's goal in this project to obtain a 10% productivity increase in its rubber processing plants if suitable water-based nanofluids can be developed and commercially produced in a cost-effective manner.

Han et al. [12] have used phase change materials as nanoparticles in nanofluids to simultaneously enhance the effective thermal conductivity and specific heat of the fluids. As an example, a suspension of indium nanoparticles (melting temperature, 157°C) in polyalphaolefin has been synthesized using a one-step, nanoemulsification method. The fluid's thermophysical properties, that is, thermal conductivity, viscosity, and specific heat, and their temperature dependence were measured experimentally. The observed melting-freezing phase transition of the indium nanoparticles significantly augmented the fluid's effective specific heat.

This work is one of the few to address thermal diffusivity; similar studies allow industrial cooling applications to continue without thorough understanding of all the heat transfer mechanisms in nanofluids.

Smart Fluids

In this new age of energy awareness, our lack of abundant sources of clean energy and the widespread dissemination of battery operated devices, such as cell-phones and laptops, have accented the necessity for a smart technological handling of energetic resources. Nanofluids have been demonstrated to be able to handle this role in some instances as a smart fluid.

In a recent paper published in the March 2009 issue of Physical Review Letters, Donzelli et al. [13] showed that a particular class of nanofluids can be used as a smart material working as a heat valve to control the flow of heat. The nanofluid can be readily configured either in a "low" state, where it conducts heat poorly, or in a "high" state, where the dissipation is more efficient. To leap the chasm to heating and cooling technologies, the researchers will have to show more evidence of a stable operating system that responds to a larger range of heat flux inputs.

Nuclear Reactors

Kim et al. [14, 15] at the Nuclear Science and Engineering Department of the Massachusetts Institute of Technology (MIT), performed a study to assess the feasibility of nanofluids in nuclear applications by improving the performance of any water-cooled nuclear system that is heat removal limited. Possible applications include pressurized water reactor (PWR) primary coolant, standby safety systems, accelerator targets, plasma divertors, and so forth, [16].

In a pressurized water reactor (PWR) nuclear power plant system, the limiting process in the generation of steam is critical heat flux (CHF) between the fuels rods and the water—when vapor bubbles that end up covering the surface of the fuel rods conduct very little heat as opposed to liquid water. Using nanofluids instead of water, the fuel rods become coated with nanoparticles such as alumina, which actually push newly formed bubbles away, preventing the formation of a layer of vapor around the rod and subsequently increasing the CHF significantly.

After testing in MIT's Nuclear Research Reactor, preliminary experiments have shown promising success where it is seen that PWR is significantly more productive. The use of nanofluids as a coolant could also be used in emergency cooling systems, where they could cool down overheat surfaces more quickly leading to an improvement in power plant safety.

Some issues regarding the use of nanofluids in a power plant system include the unpredictability of the amount of nanoparticles that are carried away by the boiling vapor. One other concern is what extra safety measures that have to be taken in the disposal of the nanofluid. The application of nanofluid coolant to boiling water reactors (BWR) is predicted to be minimal because nanoparticle carryover to the turbine and condenser would raise erosion and fouling concerns.

From Jackson's study [17], it was observed that considerable enhancement in the critical heat flux can be achieved by creating a structured surface from the deposition of nanofluids. If the deposition film characteristics such as the structure and thickness can be controlled it may be possible to increase the CHF with little decrease in the heat transfer. Whereas the nanoparticles themselves cause no significant difference in the pool-boiling characteristics of water, the boiling of nanofluids shows promise as a simple way to create an enhanced surface.

The use of nanofluids in nuclear power plants seems like a potential future application [16]. Several significant gaps in knowledge are evident at this time, including, demonstration of the nanofluid thermal-hydraulic performance at prototypical reactor conditions and the compatibility of the nanofluid chemistry with the reactor materials.

Another possible application of nanofluids in nuclear systems is the alleviation of postulated severe accidents during which the core melts and relocates to the bottom of the reactor vessel. If such accidents were to occur, it is desirable to retain the molten fuel within the vessel by removing the decay heat through the vessel wall. This process is limited by the occurrence of CHF on the vessel outer surface, but analysis indicates that the use of nanofluid can increase the in-vessel retention capabilities of nuclear reactors by as much as 40% [18].

Many water-cooled nuclear power systems are CHF-limited, but the application of nanofluid can greatly improve the CHF of the coolant so that there is a bottom-line economic benefit while also raising the safety standard of the power plant system.

Extraction of Geothermal Power and other Energy Sources

The world's total geothermal energy resources were calculated to be over 13000 ZJ in a report from MIT (2007) [19]. Currently only 200 ZJ would be extractable, however, with technological improvements, over 2,000 ZJ could be extracted and supply the world's energy needs for several millennia. When extracting energy from the earth's crust that varies in length between 5 to 10 km and temperature between and 500°C and 1000°C, nanofluids can be employed to cool the pipes exposed to such high temperatures. When drilling, nanofluids can serve in cooling the machinery and equipment working in high friction and high temperature environment. As a "fluid superconductor," nanofluids could be used as a working fluid to extract energy from the earth core and processed in a PWR power plant system producing large amounts of work energy.

In the sub-area of drilling technology, so fundamental to geothermal power, improved sensors and electronics cooled by nanofluids capable of operating at higher temperature in downhole tools, and revolutionary improvements utilizing new methods of rock penetration cooled and lubricated by nanofluids will lower production costs. Such improvements will enable access to deeper, hotter regions in high grade formations or to economically acceptable temperatures in lower-grade formations.

In the sub-area of power conversion technology, improving heat-transfer performance for lower-temperature nanofluids, and developing plant designs for higher resource temperatures to the supercritical water region would lead to an order of magnitude (or more) gain in both reservoir performance and heat-to power conversion efficiency.

Tran et al. [20], funded by the United States Department of Energy (USDOE), performed research targeted at developing a new class of highly specialized drilling fluids that may have superior performance in high temperature drilling. This research is applicable to high pressure high temperature drilling, which may be pivotal in opening up large quantities of previously unrecoverable domestic fuel resources. Commercialization would be the bottleneck of progress in this sub-area.

AUTOMOTIVE APPLICATIONS

Engine oils, automatic transmission fluids, coolants, lubricants, and other synthetic high-temperature heat transfer fluids found in conventional truck thermal systems—radiators, engines, heating, ventilation and air-conditioning (HVAC)—have inherently poor heat transfer properties. These could benefit from the high thermal conductivity offered by nanofluids that resulted from addition of nanoparticles [2, 21].

Nanofluid Coolant

In looking for ways to improve the aerodynamic designs of vehicles, and subsequently the fuel economy, manufacturers must reduce the amount of energy needed to overcome wind resistance on the road. At high speeds, approximately 65% of the total energy output from a truck is expended in overcoming the aerodynamic drag. This fact is partly due to the large radiator in front of the engine positioned to maximize the cooling effect of oncoming air.

The use of nanofluids as coolants would allow for smaller size and better positioning of the radiators. Owing to the fact that there would be less fluid due to the higher efficiency, coolant pumps could be shrunk and truck engines could be operated at higher temperatures allowing for more horsepower while still meeting stringent emission standards.

Argonne researchers, Singh et al. [22], have determined that the use of high-thermal conductive nanofluids in radiators can lead to a reduction in the frontal area of the radiator by up to 10%. This reduction in aerodynamic drag can lead to a fuel savings of up to 5%. The application of nanofluid also contributed to a reduction of friction and wear, reducing parasitic losses, operation of components such as pumps and compressors, and subsequently leading to more than 6% fuel savings. It is conceivable that greater improvement of savings could be obtained in the future.

In order to determine whether nanofluids degrade radiator material, they have built and calibrated an apparatus that can emulate the coolant flow in a radiator and are currently testing and measuring material loss of typical radiator materials by various nanofluids. Erosion of radiator material is determined by weight loss-measurements as a function of fluid velocity and impact angle.

In their tests, they observed no erosion using nanofluids made from base fluids ethylene and tri-cloroethylene gycols with velocities as high as 9 m/s and at 90°-30° impact angles. There was erosion observed with copper nanofluid at a velocity of 9.6 m/s and impact angle of 90°. The corresponding recession rate was calculated to be 0.065 mils/yr of vehicle operation.

Through preliminary investigation, it was determined that copper nanofluid produces a higher wear rate than the base fluid and this is possibly due to oxidation of copper nanoparticles. A lower wear and friction rate was seen for alumina nanofluids in comparison to the base fluid. Some interesting erosion test results from Singh et al. [22] are shown in Tables 1 and 2.

Table 1: Erosion Test Results for 50% Ethlyene Glycol, 50% H_2O Aluminum 3003 - 50°C Rig [22]

Impact Angle (*)	Velocity (m/s)	Time (hrs)	Weight Loss (mg)
90	8.0	236	0 ± 0.2
90	10.5	211	0 ± 0.2
50	6.0	264	0 ± 0.2
50	10.0	244	0 ± 0.2
30	8.0	283	0 ± 0.2
30	10.5	293	0 ± 0.2

Table 2: Erosion Test Results for Cu Nanoparticles in Tri-chloroethylene Glycol on Al 3003 - 50°C Rig [22]

Impact Angle (*)	Velocity (m/s)	Time (hrs)	Weight Loss (mg)
90	4.0	217	0 ± 0.2
30	4.0	311	0 ± 0.2
90	7.6	341	0 ± 0.2
30	7.6	335	0 ± 0.2
30	9.6	336	0 ± 0.2

Shen et al. [23] researched the wheel wear and tribological characteristics in wet, dry and minimum quantity lubrication (MQL) grinding of cast iron. Water-based alumina and diamond nanofluids were applied in the MQL grinding process and the grinding results were compared with those of pure water. Nanofluids demonstrated the benefits of reducing grinding forces, improving surface roughness, and preventing burning of the workpiece. Contrasted to dry grinding, MQL grinding could considerably lower the grinding temperature.

More research must be conducted on the tribological properties using nanofluids of a wider range of particle loadings as well as on the erosion rate of radiator material in order to help develop predictive models for nanofluid wear and erosion in engine systems.

Future research initiatives involve nanoparticles materials containing aluminum and oxide-coated metal nanoparticles. Additional research and testing in this area will assist in the design of engine cooling and other thermal management systems that involve nanofluids.

Future engines that are designed using nanofluids' cooling properties would be able to run at more optimal temperatures allowing for increased power output. With a nanofluids engine, components would be smaller and weigh less allowing for better gas mileage, saving consumers money and resulting in fewer emissions for a cleaner environment.

Nanofluid in Fuel

The aluminum nanoparticles, produced using a plasma arc system, are covered with thin layers of aluminum oxide, owing to the high oxidation activity of pure aluminum, thus creating a larger contact surface area with water and allowing for increased decomposition of hydrogen from water during the combustion process. During this combustion process, the alumina acts as a catalyst and the aluminum nanoparticles then serve to decompose the water to yield more hydrogen. It was shown that the combustion of diesel fuel mixed with aqueous aluminum nanofluid increased the total combustion heat while decreasing the concentration of smoke and nitrous oxide in the exhaust emission from the diesel engine [24, 25].

Brake and other Vehicular Nanofluids

As vehicle aerodynamics is improved and drag forces are reduced, there is a higher demand for braking systems with higher and more efficient heat dissipation mechanisms and properties such as brake nanofluid.

A vehicle's kinetic energy is dispersed through the heat produced during the process of braking and this is transmitted throughout the brake fluid in the hydraulic braking system. If the heat causes the brake fluid to reach its boiling point, a vapor-lock is created that retards the hydraulic system from dispersing the heat caused from braking. Such an occurrence will in turn will cause a brake malfunction and poses a safety hazard in vehicles. Since brake oil is easily affected by the heat generated from braking, nanofluids with enhanced characteristics maximize performance in heat transfer as well as remove any safety concerns.

Copper-oxide brake nanofluid (CBN) is manufactured using the method of arc-submerged nanoparticle synthesis system (ASNSS). Essentially this is done by melting bulk copper metal used as the electrode which is submerged in dielectric liquid within a vacuum-operating environment and the vaporized metals are condensed in the dielectric liquid [24, 25].

Aluminum-oxide brake nanofluid (AOBN) is made using the plasma charging arc system. This is performed in a very similar fashion to that of the ASNSS method. The aluminum metal is vaporized by the plasma electric arc at a high temperature and mixed thoroughly with the dielectric liquid [24, 25].

CBN has a thermal conductivity 1.6 times higher than that of the brake fluid designated DOT3, while AOBN's thermal conductivity is only 1.5 times higher than DOT3. This enhanced thermal conductivity optimizes heat transmission and lubrication.

CBN and AOBN both have enhanced properties such as a higher boiling point, higher viscosity and a higher conductivity than that of traditional brake fluid (DOT3). By yielding a higher boiling point, conductivity and viscosity, CBN and AOBN reduce the occurrence of vapor-lock and offer increased safety while driving. Important findings of Kao et al. [24, 25] are shown in Figure 1 and Table 3.

Table 3: CBN and AOBN Boiling Point and Thermal Conductivity Fluctuations [24, 25]

	DOT3*	CBN 2 wt% (CuO + DOT3)	DOT3*	AOBN 2 wt% (Al$_2$O$_3$ + DOT3)
Boiling Point	270°C	278°C	240°C	248°C
Conductivity (25°C)	0.03 W/m°C	0.05 W/m°C	0.13 W/m°C	0.19 W/m°C

*Different DOT3 brake fluids were used.

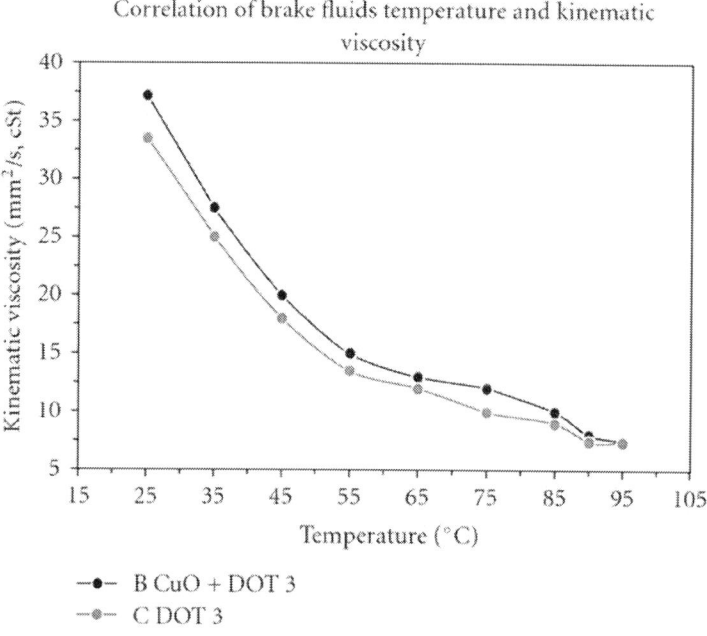

Figure 1: CBN Temperature and Viscosity Fluctuations [24, 25].

In the nanofluid research applied to the cooling of automatic transmissions, Tzeng et al. [26] dispersed CuO and Al_2O_3 nanoparticles into engine transmission oil. The experimental setup was the transmission of a four-wheel-drive vehicle. The transmission had an advanced rotary blade coupling, where high local temperatures occurred at high rotating speeds. Temperature measurements were taken on the exterior of the rotary-blade-coupling transmission at four engine operating speeds (range from 400 to 1600 rpm), and the optimum composition of nanofluids with regard to heat transfer performance was studied. The results indicated that CuO nanofluids resulted in the lowest transmission temperatures both at high and low rotating speeds. Therefore, the use of nanofluid in the transmission has a clear advantage from the thermal performance viewpoint. As in all nanofluid applications, however, consideration must be given to such factors as particle settling, particle agglomeration, and surface erosion.

In automotive lubrication applications, Zhang [27] reported that surface-modified nanoparticles stably dispersed in mineral oils are effective in reducing wear and enhancing load-carrying capacity. Results from a research project involving industry and academia points to the use of nanoparticles in lubricants to enhance tribological properties such as load-carrying capacity, wear resistance, and friction reduction between moving mechanical components. Such results are promising for enhancing heat transfer rates in automotive systems through the use of nanofluids.

ELECTRONIC APPLICATIONS

Nanofluids are used for cooling of microchips in computers and elsewhere. They are also used in other electronic applications which use microfluidic applications.

Cooling of Microchips

A principal limitation on developing smaller microchips is the rapid heat dissipation. However, nanofluids can be used for liquid cooling of computer processors due to their high thermal conductivity. It is predicted that the next generation of computer chips will produce localized heat flux over $10\,MW/m^2$, with the total power exceeding $300\,W$. In combination with thin film evaporation, the nanofluid oscillating heat pipe (OHP) cooling system will be able to remove heat fluxes over $10\,MW/m^2$ and serve as the next generation cooling device that will be able to handle the heat dissipation coming from new technology [28, 29].

In order to observe the oscillation, researchers had to modify the metal pipe system of the OHP to use glass or plastic for visibility. However, since OHP systems are usually made of copper, the use of glass or plastic changes the thermal transfer properties of the system and subsequently altering the performance of the system and the legitimacy of the experimental data [28, 29].

So as to obtain experimental data while maintaining the integrity of the OHP system, Arif [30] employed neutron imaging to study the liquid flow in a 12-turn nanofluid OHP. As a consequence of the high intensity neutron beam from an amorphous silicon imaging system, they were able to capture dynamic images at 1/30th of a second. The nanofluid used was composed of diamond nanoparticles suspended in water.

Even though nanofluids and OHPs are not new discoveries, combining their unique features allows for the nanoparticles to be completely suspended in the base liquid increasing their heat transport capability. Since nanofluids have a strong temperature-dependent thermal conductivity and they show a nonlinear relationship between thermal conductivity and concentration, they are high performance conductors with an increased CHF. The OHP takes intense heat from a high-power device and converts it into kinetic energy of fluids while not allowing the liquid and vapor phases to interfere with each other since they flow in the same direction.

In their experiment, Ma et al. [28, 29] introduced diamond nanoparticles into high performance liquid chromatography (HPLC) water. The movement of the OHP keeps the nanoparticles from settling and thus improving the efficiency of the cooling device. At an input power of 80 W, the diamond nanofluid decreased the temperature difference between the evaporator and the condenser from to 40.9°C to 24.3°C.

However, as the heat input increases, the oscillating motion increases and the resultant temperature difference between the evaporator and condenser does not continue to increase after a certain power input. This phenonmenon inhibits the effective thermal conductivity of the nanofluid from continuously increasing. However, at its maximum power level of 336 W, the temperature difference for the nanofluid OHP was still less than that for the OHP with pure water, Figure 2. Hence, it has been shown that the nanofluid can significantly increase the heat transport capability of the OHP.

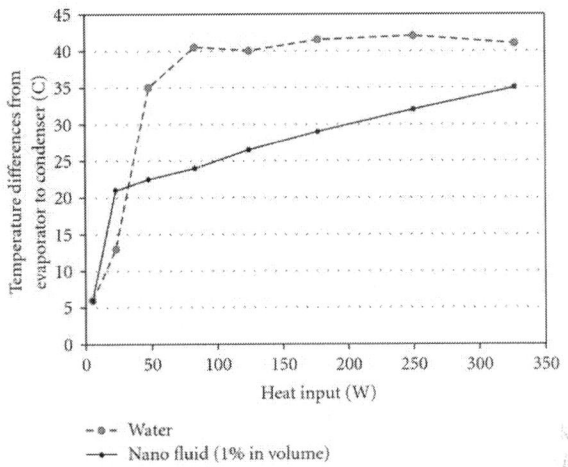

Figure 2: Effect of nanofluid on heat transport capability in an OHP [28, 29].

Lin et al. [31] investigated nanofluids in pulsating heat pipes by using silver nanoparticles, and discovered encouraging results. The silver nanofluid improved heat transfer characteristics of the heat pipes.

Nguyen et al. [32] investigated the heat transfer enhancement and behavior of Al_2O_3-water nanofluid with the intention of using it in a closed cooling system designed for microprocessors or other electronic devices. The experimental data supports that the inclusion of nanoparticles into distilled water produces a significant increase of the cooling convective heat transfer coefficient. At a given particle concentration of 6.8%, the heat transfer coefficient increased as much as 40% compared to the base fluid of water. Smaller Al_2O_3 nanoparticles also showed higher convective heat transfer coefficients than the larger ones.

Further research of nanofluids in electronic cooling applications will lead to the development of the next generation of cooling devices that incorporate nanofluids for ultrahigh-heat-flux electronic systems.

Microscale Fluidic Applications

The manipulation of small volumes of liquid is necessary in fluidic digital display devices, optical devices, and microelectromechanical systems (MEMS) such as lab-on-chip analysis systems. This can be done by electrowetting, or reducing the contact angle by an applied voltage, the small volumes of liquid. Electrowetting on dielectric (EWOD) actuation is one very useful method of microscale liquid manipulation.

Vafaei et al. [33] discovered that nanofluids are effective in engineering the wettability of the surface and possibly of surface tension. Using a goniometer, it was observed that even the addition of a very low concentration of bismuth telluride nanofluid dramatically changed the wetting characteristics of the surface. Concentrations as low as 3×10^{-6} increased the contact angle to over 40°C, distinctly indicating that the nanoparticles change the force balance in the vicinity of the triple line. The contact angle, q° Figure 3 rises with the concentration of the nanofluid, reaches a maximum, and then decreases, Figure 4 [34].

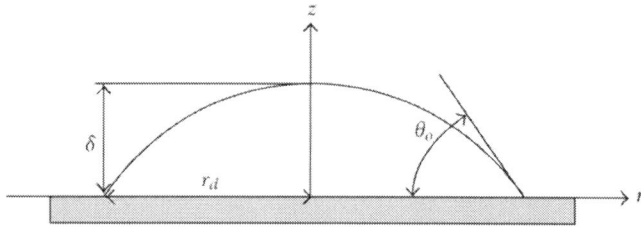

Figure 3: Schematic diagram of droplet shape.

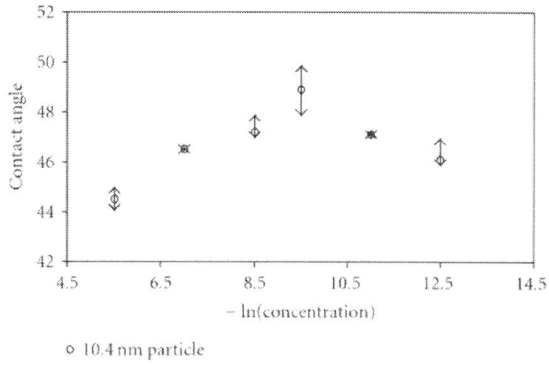

Figure 4: Variation of contact angle for 10 nm bismuth telluride nanoparticles concentration on a glass substrate [34].

The droplet contact angle was observed to change depending on the size of the nanoparticles as well. Smaller nanoparticles are more effective in increasing the contact angle. The reason for this effect is that smaller particles would provide more surface-to-volume area, for the same concentration.

Dash et al. used the EWOD effect to demonstrate that nanofluids display increased performance and stability when exposed to electric fields, Figure 5. The experiment consisted of placing droplets of water-based solutions containing bismuth telluride nanoparticles onto a Teflon-coated silicon wafer. A strong change in the angle at with the droplet contacted the wafer was observed when an electric field was applied to the droplet. The change noticed with the nanofluids was significantly greater than when not using nanofluids. The bismuth telluride nanofluid also displayed enhanced droplet stability and absence of the contact angle saturation effect compared to solutions of 0.01 N Na_2SO_4 and thioglycolic acid in deionized water.

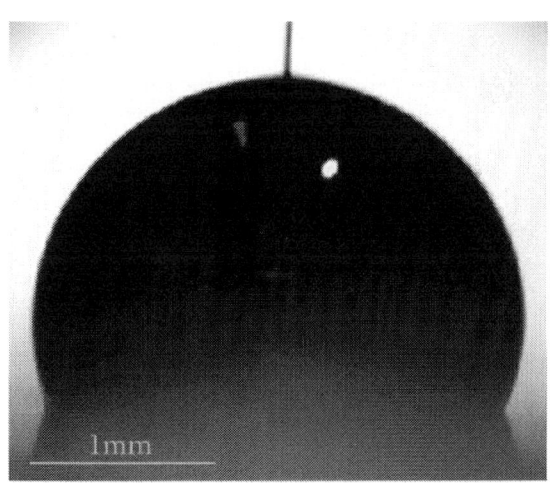

(a)

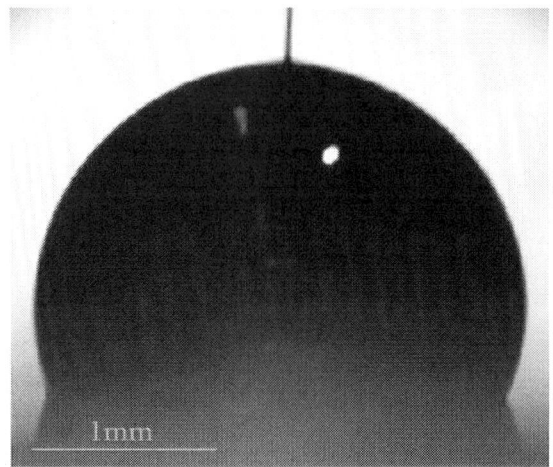

(b)

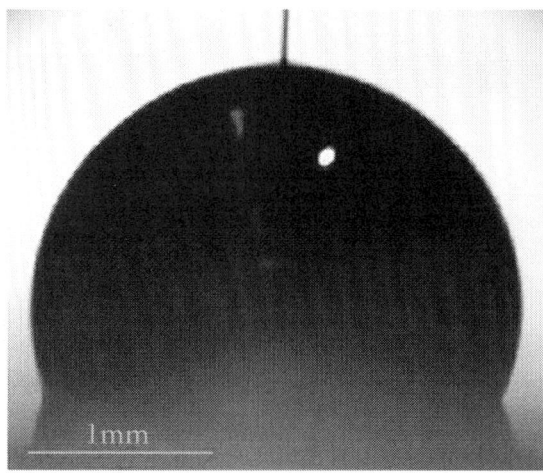

(c)

Applications of Nanofluids: Current and Future

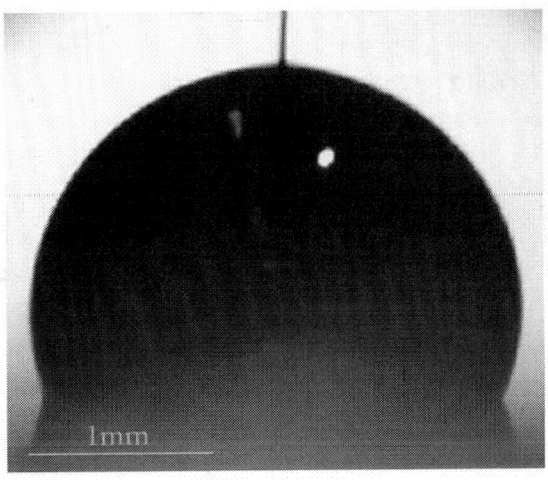

(d)

Figure 5: Nanofluid droplet with applied voltage (a) 0 V and (b) 64.5 V 0.01 M droplet under (c) 0 V and (d) 40 V. The Na_2SO_4 droplet becomes unstable at approximately 40 V seen by the gas bubbles forming in the contact line, [34].

That the contact angle of droplets of nanofluids can be changed has potential applications for efficiently moving liquids in microsystems, allowing for new methods for focusing lenses in miniature cameras as well as for cooling computer chips.

BIOMEDICAL APPLICATIONS

Nanodrug Delivery

Most bio-MEMS studies were done in academia in the 1990s, while recently commercializations of such devices have started. Examples include an electronically activated drug delivery microchip [35]; a controlled delivery system via integration of silicon and electroactive polymer technologies; a MEMS-based DNA sequencer developed by Cepheid [36]; and arrays of in-plane and out-of-plane hollow microneedles for dermal/transdermal drug delivery [37, 38] as well as nanomedicine applications of nanogels or gold-coated nanoparticles [39]. An objective of the advanced endeavors in developing integrated micro- or nano-drug delivery systems is the interest in easily monitoring and controlling target-cell responses to pharmaceutical stimuli, to understand biological cell activities, or to enable drug development processes.

While conventional drug delivery is characterized by the "high-and-low" phenomenon, microdevices facilitate precise drug delivery by both implanted and transdermal techniques. This means that when a drug is dispensed conventionally, drug concentration in the blood will increase, peak and then drop as the drug is metabolized, and the cycle is repeated for each drug dose. Employing nano-drug delivery (ND) systems, controlled drug release takes place over an extended period of time. Thus, the desired drug concentration will be sustained within the therapeutic window as required.

A nanodrug-supply system, that is, a bio-MEMS, was introduced by Kleinstreuer et al. [40]. Their principal concern were the conditions for delivering uniform concentrations at the microchannel exit of the supplied nano-drugs. A heat flux which depends on the levels of nano-fluid and purging fluid velocity was added to ascertain that drug delivery to the living cells occurs at an optimal temperature, that is, 37°C. The added wall heat flux had also a positive influence on drug-concentration uniformity. In general, the nano-drug concentration uniformity is affected by channel length, particle diameter and the Reynolds number of both the nanofluid supply and main microchannels. Since the transport mechanisms are dependent on convection— diffusion, longer channels, smaller particle diameters as well as lower Reynolds numbers are desirable for best, that is, uniform drug delivery.

Cancer Theraupetics

There is a new initiative which takes advantage of several properties of certain nanofluids to use in cancer imaging and drug delivery. This initiative involves the use of iron-based nanoparticles as delivery vehicles for drugs or radiation in cancer patients. Magnetic nanofluids are to be used to guide the particles up the bloodstream to a tumor with magnets. It will allow doctors to deliver high local doses of drugs or radiation without damaging nearby healthy tissue, which is a significant side effect of traditional cancer treatment methods. In addition, magnetic nanoparticles are more adhesive to tumor cells than non-malignant cells and they absorb much more power than microparticles in alternating current magnetic fields tolerable in humans; they make excellent candidates for cancer therapy.

Magnetic nanoparticles are used because as compared to other metal-type nanoparticles, these provide a characteristic for handling and manipulation of the nanofluid by magnetic force [41]. This combination of targeted delivery and controlled release will also decrease the likelihood of systemic toxicity since the drug is encapsulated and biologically unavailable during transit in systemic circulation. The nanofluid containing magnetic nanoparticles also acts as a super-paramagnetic fluid which in an alternating electromagnetic field absorbs energy producing a controllable hyperthermia. By enhancing the chemotherapeutic efficacy, the hyperthermia is able to produce a preferential radiation effect on malignant cells [42].

There are numerous biomedical applications that involve nanofluids such as magnetic cell separation, drug delivery, hyperthermia, and contrast enhancement in magnetic resonance imaging. Depending on the specific application, there are different chemical syntheses developed for various types of magnetic nanofluids that allow for the careful tailoring of their properties for different requirements in applications. Surface coating of nanoparticles and the colloidal stability of biocompatible water-based magnetic fluids are the two particularly important factors that affect successful application [43, 44].

Nanofluids could be applied to almost any disease treatment techniques by reengineering the nanoparticles' properties. In their study, the nanoparticles were laced with the drug docetaxel to be dissolved in the cells' internal fluids, releasing the anticancer drug at a predetermined rate. The nanoparticles contain targeting molecules called aptamers which recognize the surface molecules on cancer cells preventing the nanoparticles from attacking other cells. In order to prevent the nanoparticles from being destroyed by macrophages—cells that guard against foreign substances entering our bodies—the nanoparticles also have polyethylene glycol molecules. The nanoparticles are excellent drug-delivery vehicles because they are so small that living cells absorb them when they arrive at the cells' surface.

For most biomedical uses the magnetic nanoparticles should be below 15 nm in size and stably dispersed in water. A potential magnetic nanofluid that could be used for biomedical applications is one composed of FePt nanoparticles. This FePt nanofluid possesses an intrinsic chemical stability and a higher saturation magnetization making it ideal for biomedical applications. However, before magnetic nanofluids can be used as drug delivery systems, more research must be conducted on the nanoparticles containing the actual drugs and the release mechanism.

Cryopreservation

Conventional cryopreservation protocols for slow-freezing or vitrification involve cell injury due to ice formation/cell dehydration or toxicity of high cryoprotectant (CPA) concentrations, respectively. In the study by X. He et al. [45], they developed a novel cryopreservation technique to achieve ultra-fast cooling rates using a quartz microcapillary (QMC). The QMC enabled vitrification of murine embryonic stem (ES) cells using an intracellular cryoprotectant concentration in the range used for slowing freezing (1–2 M). More than 70% of the murine ES cells post-vitrification attached with respect to non-frozen control cells, and the proliferation rates of the two groups were alike. Preservation of undifferentiated properties of the pluripotent murine ES cells post-vitrification cryopreservation was verified using three different types of assays. These results indicate that vitrification at a low concentration (2 M) of intracellular cryoprotectants is a viable and effective approach for the cryopreservation of murine embryonic stem cells.

Nanocryosurgery

Cryosurgery is a procedure that uses freezing to destroy undesired tissues. This therapy is becoming popular because of its important clinical advantages. Although it still cannot be regarded as a routine method of cancer treatment, cryosurgery is quickly becoming as an alternative to traditional therapies.

Simulations were performed by Yan and Liu [46] on the combined phase change bioheat transfer problems in a single cell level and its surrounding tissues, to explicate the difference of transient temperature response between conventional cyrosugery and nanocyrosurgery. According to theoretical interpretation and existing experimental measurements, intentional loading of nanoparticles with high thermal conductivity into the target tissues can reduce the final temperature, increase the maximum freezing rate, and enlarge the ice volume obtained in the absence of nanoparticles. Additionally, introduction of nanoparticle enhanced freezing could also make conventional cyrosurgery more flexible in many aspects such as artificially interfering in the size, shape, image and direction of iceball formation. The concepts of nanocyrosurgery may offer new opportunities for future tumor treatment.

With respect to the choice of particle for enhancing freezing, magnetite (Fe_3O_4) and diamond are perhaps the most popular and appropriate because of their good biological compatibility. Particle sizes less than 10 µm are sufficiently small to start permitting effective delivery to the site of the tumor, either via encapsulation in a larger moiety or suspension in a carrier fluid. Introduction of nanoparticles into the target via a nanofluid would effectively increase the nucleation rate at a high temperature threshold.

Sensing and Imaging

Colloidal gold has been used for several centuries now, be it as colorant of glass ("Purple of Cassius") and silk, in medieval medicine for the diagnosis of syphilis or, more recently, in chemical catalysis, non-linear optics, supramolecular chemistry, molecular recognition and the biosciences. Colloidal gold is often referred to as the most stable of all colloids. Its history, properties and applications have been reviewed extensively. For a thorough and up-to-date overview the paper by Daniel and Astruc [48] and the references cited therein may be consulted. As stated in the introduction, no attempt is made here to review the use of colloids which are also nanofluids. An increase of colloids which are nanofluids is expected in this category.

OTHER APPLICATIONS

Nanofluid Detergent

Nanofluids do not behave in the same manner as simple liquids with classical concepts of spreading and adhesion on solid surfaces [7, 49, 50]. This fact opens up the possibility of nanofluids being excellent candidates in the processes of soil remediation, lubrication, oil recovery and detergency. Future engineering applications could abound in such processes.

Wasan and Nikolov [47] of Illinois Institute of Technology in Chicago were able to use reflected-light digital video microscopy to determine the mechanism of spreading dynamics in liquid containing nanosized polystyrene particles, Figure 6. They were able to demonstrate the two-dimensional crystal-like formation of the polystyrene spheres in water and how this enhances the spreading dynamics of a micellar fluid at the three-phase region [47].

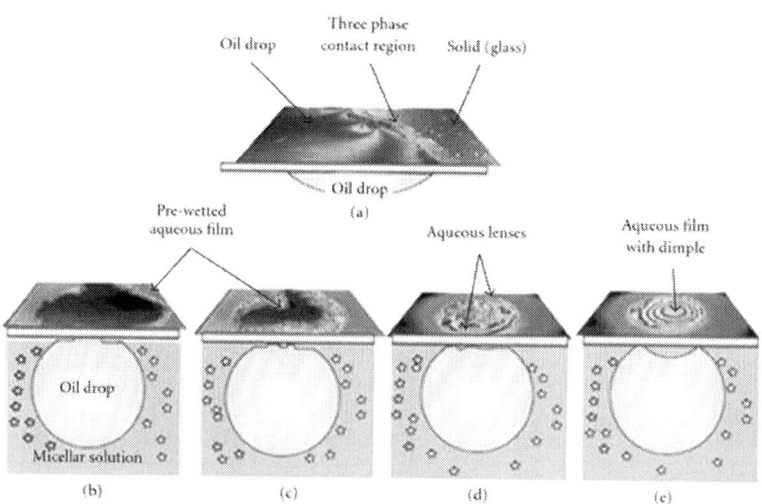

Figure 6: (a) Photomicrograph showing the oil drop placed on a glass surface and differential interference patterns formed at the three-phase contact re-

gion [47]. (b), Photomicrographs taken after addition of the nanofluid at (b), 30 s; 2 minutes; (d), 4 minutes; (e), 6 minutes region [47].

When encountering an oil drop, the polystyrene nanoparticles concentrate and rearrange around the drop creating a wedge-like region between the surface and the oil drop. The nanoparticles then diffuse into the wedge film and cause an increase in concentration and subsequently an increase in disjoining pressure around the film region. Owing to the increase in pressure, the oil-solution interface moves forward allowing the polystyrene nanoparticles to spread along the surface. It is this mechanism that causes the oil drop to detach completely from the surface.

Wasan and Nikolov [47] performed an additional experiment where they introduced an electrolyte into the process in order to decrease the interfacial tension at the interface of the oil and the nanofluid, but found that the drop did not become detached from the surface. They actually observed a diminished disjoining pressure contrary to the logical prediction. Additional work must be done in this area to determine such behavior of the nanofluid.

Overall, this phenomenon which involves the increased spreading of the detergent surfactants, which are not only limited to polystyrene nanoparticles, and enhanced oil removal process offers a new way of removing stains and grease from surfaces. This type of nanofluid also has potential in the commercial extraction of oil from the ground as well as the remediation of oil spills.

CONCLUSIONS

Nanofluids are important because they can be used in numerous applications involving heat transfer, and other applications such as in detergency. Colloids which are also nanofluids have been used in the biomedical field for a long time, and their use will continue to grow. Nanofluids have also been demonstrated for use as smart fluids. Problems of nanoparticle agglomeration, settling, and erosion potential all need to be examined in detail in the applications. Nanofluids employed in experimental research have to be well characterized with respect to particle size, size distribution, shape and clustering so as to render the results most widely applicable. Once the science and engineering of nanofluids are fully understood and their full potential researched, they can be reproduced on a large scale and used in many applications. Colloids which are also nanofluids will see an increase in use in biomedical engineering and the biosciences.

Further research still has to be done on the synthesis and applications of nanofluids so that they may be applied as predicted. Nevertheless, there have been many discoveries and improvements identified about the characteristics of nanofluids in the surveyed applications and we are a step closer to developing systems that are more efficient and smaller, thus rendering the environment cleaner and healthier.

REFERENCES

1. S. U. S. Choi, "Nanofluids: from vision to reality through research," Journal of Heat Transfer, vol. 131, no. 3, pp. 1–9, 2009.
2. W. Yu, D. M. France, J. L. Routbort, and S. U. S. Choi, "Review and comparison of nanofluid thermal conductivity and heat transfer enhancements," Heat Transfer Engineering, vol. 29, no. 5, pp. 432–460, 2008. ·
3. T. Tyler, O. Shenderova, G. Cunningham, J. Walsh, J. Drobnik, and G. McGuire, "Thermal transport properties of diamond-based nanofluids and nanocomposites," Diamond and Related Materials, vol. 15, no. 11-12, pp. 2078–2081, 2006. ·
4. S. K. Das, S. U. S. Choi, and H. E. Patel, "Heat transfer in nanofluids—a review," Heat Transfer Engineering, vol. 27, no. 10, pp. 3–19, 2006. ·
5. M.-S. Liu, M. C.-C. Lin, I.-T. Huang, and C.-C. Wang, "Enhancement of thermal conductivity with carbon nanotube for nanofluids," International Communications in Heat and Mass Transfer, vol. 32, no. 9, pp. 1202–1210, 2005. ·
6. S. U. S. Choi, Z. G. Zhang, and P. Keblinski, "Nanofluids," in Encyclopedia of Nanoscience and Nanotechnology, H. S. Nalwa, Ed., vol. 6, pp. 757–737, American Scientific, Los Angeles, Calif, USA, 2004.
7. S. M. S. Murshed, S.-H. Tan, and N.-T. Nguyen, "Temperature dependence of interfacial properties and viscosity of nanofluids

for droplet-based microfluidics," Journal of Physics D, vol. 41, no. 8, Article ID 085502, 5 pages, 2008. ·

8. K.-F. V. Wong and T. Kurma, "Transport properties of alumina nanofluids," Nanotechnology, vol. 19, no. 34, Article ID 345702, 8 pages, 2008. ·

9. K.-F. V. Wong, B. L. Bon, S. Vu, and S. Samedi, "Study of nanofluid natural convection phenomena in rectangular enclosures," in Proceedings of the ASME International Mechanical Engineering Congress and Exposition (IMECE ‹07), vol. 6, pp. 3–13, Seattle, Wash, USA, November 2007.

10. Y. Ju-Nam and J. R. Lead, "Manufactured nanoparticles: an overview of their chemistry, interactions and potential environmental implications," Science of the Total Environment, vol. 400, no. 1–3, pp. 396–414, 2008. · ·

11. J. Routbort, et al., Argonne National Lab, Michellin North America, St. Gobain Corp., 2009,http://www1.eere.energy.gov/industry/nanomanufacturing/pdfs/nanofluids_industrial_cooling.pdf.

12. Z. H. Han, F. Y. Cao, and B. Yang, "Synthesis and thermal characterization of phase-changeable indium/polyalphaolefin nanofluids," Applied Physics Letters, vol. 92, no. 24, Article ID 243104, 3 pages, 2008. ·

13. G. Donzelli, R. Cerbino, and A. Vailati, "Bistable heat transfer in a nanofluid," Physical Review Letters, vol. 102, no. 10, Article ID 104503, 4 pages, 2009. ·

14. S. J. Kim, I. C. Bang, J. Buongiorno, and L. W. Hu, "Study of pool boiling and critical heat flux enhancement in nanofluids," Bulletin of the Polish Academy of Sciences—Technical Sciences, vol. 55, no. 2, pp. 211–216, 2007.

15. S. J. Kim, I. C. Bang, J. Buongiorno, and L. W. Hu, "Surface wettability change during pool boiling of nanofluids and its effect on critical heat flux," International Journal of Heat and Mass Transfer, vol. 50, no. 19-20, pp. 4105–4116, 2007. ·

16. J. Boungiorno, L.-W. Hu, S. J. Kim, R. Hannink, B. Truong, and E. Forrest, "Nanofluids for enhanced economics and safety of nuclear reactors: an evaluation of the potential features issues, and research gaps," Nuclear Technology, vol. 162, no. 1, pp. 80–91, 2008.

17. E. Jackson, Investigation into the pool-boiling characteristics of gold nanofluids, M.S. thesis, University of Missouri-Columbia, Columbia, Mo, USA, 2007.
18. J. Buongiorno, L. W. Hu, G. Apostolakis, R. Hannink, T. Lucas, and A. Chupin, "A feasibility assessment of the use of nanofluids to enhance the in-vessel retention capability in light-water reactors," Nuclear Engineering and Design, vol. 239, no. 5, pp. 941–948, 2009. ·
19. "The Future of Geothermal Energy," MIT, Cambridge, Mass, USA, 2007.
20. P. X. Tran, D. K. lyons, et al., "Nanofluids for Use as Ultra-Deep Drilling Fluids," U.S.D.O.E., 2007,http://www.netl.doe.gov/publications/factsheets/rd/R&D108.pdf.
21. M. Chopkar, P. K. Das, and I. Manna, "Synthesis and characterization of nanofluid for advanced heat transfer applications," Scripta Materialia, vol. 55, no. 6, pp. 549–552, 2006. ·
22. D. Singh, J. Toutbort, G. Chen, et al., "Heavy vehicle systems optimization merit review and peer evaluation," Annual Report, Argonne National Laboratory, 2006.
23. B. Shen, A. J. Shih, S. C. Tung, and M. Hunter, "Application of nanofluids in minimum quantity lubrication grinding," Tribology and Lubrication Technology.
24. M. J. Kao, C. H. Lo, T. T. Tsung, Y. Y. Wu, C. S. Jwo, and H. M. Lin, "Copper-oxide brake nanofluid manufactured using arc-submerged nanoparticle synthesis system," Journal of Alloys and Compounds, vol. 434-435, pp. 672–674, 2007. ·
25. M. J. Kao, H. Chang, Y. Y. Wu, T. T. Tsung, and H. M. Lin, "Producing aluminum-oxide brake nanofluids using plasma charging system," Journal of the Chinese Society of Mechanical Engineers, vol. 28, no. 2, pp. 123–131, 2007.
26. S.-C. Tzeng, C.-W. Lin, and K. D. Huang, "Heat transfer enhancement of nanofluids in rotary blade coupling of four-wheel-drive vehicles," Acta Mechanica, vol. 179, no. 1-2, pp. 11–23, 2005. ·
27. Q. Xue, J. Zhang, and Z. Zhang, "Synthesis, structure and lubricating properties of dialkyldithiophosphate-modified Mo-S

compound nanoclusters," Wear, vol. 209, no. 1-2, pp. 8–12, 1997.

28. H. B. Ma, C. Wilson, B. Borgmeyer, et al., "Effect of nanofluid on the heat transport capability in an oscillating heat pipe," Applied Physics Letters, vol. 88, no. 14, Article ID 143116, 3 pages, 2006.

29. H. B. Ma, C. Wilson, Q. Yu, K. Park, U. S. Choi, and M. Tirumala, "An experimental investigation of heat transport capability in a nanofluid oscillating heat pipe," Journal of Heat Transfer, vol. 128, no. 11, pp. 1213–1216, 2006. ·

30. M. Arif, "Neutron imaging for fuel cell research," in Proceedings of the Imaging and Neutron Workshop, Oak Ridge, Tenn, USA, October 2006.

31. Y.-H. Lin, S.-W. Kang, and H.-L. Chen, "Effect of silver nano-fluid on pulsating heat pipe thermal performance," Applied Thermal Engineering, vol. 28, no. 11-12, pp. 1312–1317, 2008. ·

32. C. T. Nguyen, G. Roy, C. Gauthier, and N. Galanis, "Heat transfer enhancement using Al2O3-water nanofluid for an electronic liquid cooling system," Applied Thermal Engineering, vol. 27, no. 8-9, pp. 1501–1506, 2007. ·

33. S. Vafaei, T. Borca-Tasciuc, M. Z. Podowski, A. Purkayastha, G. Ramanath, and P. M. Ajayan, "Effect of nanoparticles on sessile droplet contact angle," Nanotechnology, vol. 17, no. 10, pp. 2523–2527, 2006. ·

34. R. K. Dash, T. Borca-Tasciuc, A. Purkayastha, and G. Ramanath, "Electrowetting on dielectric-actuation of microdroplets of aqueous bismuth telluride nanoparticle suspensions," Nanotechnology, vol. 18, no. 47, Article ID 475711, 6 pages, 2007. ·

35. R. S. Shawgo, A. C. R. Grayson, Y. Li, and M. J. Cima, "BioMEMS for drug delivery," Current Opinion in Solid State and Materials Science, vol. 6, no. 4, pp. 329–334, 2002. · ·

36. Cepheid, 2009, http://www.Cepheid.Com.

37. A. Ovsianikov, B. Chichkov, P. Mente, N. A. Monteiro-Riviere, A. Doraiswamy, and R. J. Narayan, "Two photon polymerization of polymer-ceramic hybrid materials for transdermal drug delivery,"International Journal of Applied Ceramic Technology, vol. 4, no. 1, pp. 22–29, 2007. ·

38. K. Kim and J.-B. Lee, "High aspect ratio tapered hollow metallic microneedle arrays with microfluidic interconnector," Microsystem Technologies, vol. 13, no. 3-4, pp. 231–235, 2007.
39. V. Labhasetwar and D. L. Leslie-Pelecky, Biomedical Applications of Nanotechnology, John Wiley & Sons, New York, NY, USA, 2007.
40. C. Kleinstreuer, J. Li, and J. Koo, "Microfluidics of nano-drug delivery," International Journal of Heat and Mass Transfer, vol. 51, no. 23-24, pp. 5590–5597, 2008. ·
41. D. Bica, L. Vékás, M. V. Avdeev, et al., "Sterically stabilized water based magnetic fluids: synthesis, structure and properties," Journal of Magnetism and Magnetic Materials, vol. 311, no. 1, pp. 17–21, 2007. ·
42. P.-C. Chiang, D.-S. Hung, J.-W. Wang, C.-S. Ho, and Y.-D. Yao, "Engineering water-dispersible FePt nanoparticles for biomedical applications," IEEE Transactions on Magnetics, vol. 43, no. 6, pp. 2445–2447, 2007. ·
43. L. Vékás, D. Bica, and M. V. Avdeev, "Magnetic nanoparticles and concentrated magnetic nanofluids: synthesis, properties and some applications," China Particuology, vol. 5, no. 1-2, pp. 43–49, 2007. ·
44. L. Vékás, D. Bica, and O. Marinica, "Magnetic nanofluids stabilized with various chain length surfactants," Romanian Reports in Physics, vol. 58, no. 3, pp. 257–267, 2006.
45. X. He, E. Y. H. Park, A. Fowler, M. L. Yarmush, and M. Toner, "Vitrification by ultra-fast cooling at a low concentration of cryoprotectants in a quartz micro-capillary: a study using murine embryonic stem cells," Cryobiology, vol. 56, no. 3, pp. 223–232, 2008. · ·
46. J.-F. Yan and J. Liu, "Nanocryosurgery and its mechanisms for enhancing freezing efficiency of tumor tissues," Nanomedicine, vol. 4, no. 1, pp. 79–87, 2008. · ·
47. D. T. Wasan and A. D. Nikolov, "Spreading of nanofluids on solids," Nature, vol. 423, no. 6936, pp. 156–159, 2003. · ·
48. M.-C. Daniel and D. Astruc, "Gold nanoparticles: assembly, supramolecular chemistry, quantum-size-related properties, and

applications toward biology, catalysis, and nanotechnology," Chemical Reviews, vol. 104, no. 1, pp. 293–346, 2004. ··

49. K. Sefiane, J. Skilling, and J. MacGillivray, "Contact line motion and dynamic wetting of nanofluid solutions," Advances in Colloid and Interface Science, vol. 138, no. 2, pp. 101–120, 2008. ··

50. Y. H. Jeong, W. J. Chang, and S. H. Chang, "Wettability of heated surfaces under pool boiling using surfactant solutions and nanofluids," International Journal of Heat and Mass Transfer, vol. 51, no. 11-12, pp. 3025–3031, 2008. ·

Towards the Computational Design of Solid Catalysts

J. K. Nørskov[1], T. Bligaard[1], J. Rossmeisl[1], and C. H. Christensen[2]

[1]Center for Atomic-scale Materials Design, Department of Physics, Building 311, Technical University of Denmark, DK-2800 Kgs. Lyngby, Denmark

[2]Haldor Topsøe A/S, Nymøllevej 55, DK-2800 Kgs. Lyngby, Denmark

ABSTRACT

Over the past decade the theoretical description of surface reactions has undergone a radical development. Advances in density functional theory mean it is now possible to describe catalytic reactions at surfaces with the detail and accuracy required for computational results to compare favourably with experiments. Theoretical methods can be used to describe surface chemical reactions in detail and to understand variations in catalytic activity from one catalyst to another. Here, we review the first steps towards using computational methods to design new catalysts. Examples include screening for catalysts with increased activity and catalysts with improved selectivity. We discuss how, in the future, such methods may be used to engineer the electronic structure of the active surface by changing its composition and structure.

INTRODUCTION

During the past century chemists have developed efficient chemical reactions for converting fossil resources into a broad range of fuels and chemicals, and this can be considered one of the most important and far-reaching scientific developments up to now. Today, essentially all transportation fuels are refined in a number of catalytic processes and most chemicals are also produced using technologies based on catalysis[1]. A few well-known examples illustrate the impact: about half of all petrol in the world is now produced by fluid catalytic cracking using specially designed zeolite catalysts, and the Haber–Bosch process — featuring an iron catalyst — continues to have a key role in the production of fertilizers. The list of important small- and large-scale processes by which fossil resources are converted into fuels and chemicals is almost endless.

Such catalytic technologies have also resulted in various environmental issues — even the best processes do not allow a complete elimination of undesirable byproducts. Many innovative, catalytic technologies have subsequently been implemented to remedy these new problems; one famous example is the precious-metal-based three-way catalyst installed in most petrol-fuelled passenger cars. Moreover, these developments have contributed to an increased use of fossil resources and thus to the increasing carbon dioxide levels in the atmosphere. Currently, there is a significant drive to relinquish our dependence on fossil fuels and to minimize the emission of carbon dioxide. Clearly, this calls for many new and improved catalytic processes, and for catalytic technologies that focus on prevention rather than on remediation.

Reducing environmental impact will require entirely new catalysts: catalysts for new processes, more active and more selective catalysts and preferably catalysts that are made from earth-abundant elements. This represents a formidable challenge and it will demand an ability to design new catalytic materials well beyond our present capabilities. The ultimate goal is to have enough knowledge of the factors determining catalytic activity to be able to tailor catalysts atom-by-atom. The catalytic properties of a material are in principle determined completely by its electronic structure, so the objective is the engineering of electronic structure by changing composition and physical structure. The approach is illustrated in Fig. 1. Over the past few decades our understanding of why particular materials are good catalysts for given reactions has improved. The challenge is to invert this problem; given that we need to catalyse a certain reaction under a set of specified conditions, which material should we choose?

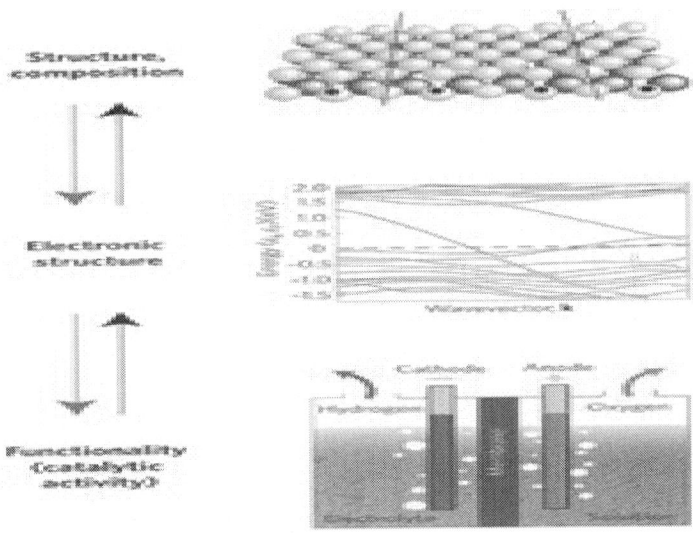

Figure 1: Tailoring materials. Illustration of the way the electronic structure is the link between the structure and composition of a material and its functionality. Changing the functionality can be achieved by engineering the electronic structure through modification of structure and composition. The example shown is a MoS_2 sheet, a few atoms wide, where new electronic states at the edges cross the Fermi level and give rise to catalytic activity, for instance in electrochemical hydrogen evolution[63].

The aim of controlling matter at the molecular scale by engineering the electronic structure is not restricted to catalytic materials; it is a general challenge in chemistry, physics and materials science, and there is considerable progress in several areas such as materials for batteries[2], hydrogen storage[3], optical absorption[4] and molecules for homogeneous catalysis[5, 6]. Catalysis at surfaces is particularly well suited for electronic structure engineering, primarily because the link between the atomic-scale properties and the macroscopic functionality — the kinetics — is well developed. In addition, the theoretical description of surface reactions has been enhanced considerably by the availability of a large number of quantitative experimental surface-science studies of adsorption and reaction phenomena[7, 8,9, 10, 11, 12].

Here, we review some of the first examples of the computer-based design of solid catalysts. We introduce a number of concepts linking catalytic performance to the properties of the catalyst's surface, and in turn discuss how the surface electronic structure determines the catalytic properties. Finally, we discuss some of the challenges ahead.

Trends and Descriptors of Catalytic Activity

The extraordinary progress in density functional theory (DFT) calculations for surface processes is the key development that has created the possibility of computer-based catalyst design[13]. Current methods are fast enough to allow the treatment of complex, extended systems[14, 15]. They can also now provide the interaction energies of molecules and atoms with metal surfaces with sufficient accuracy to describe trends in reactivity for transition metals and alloys[16].

There are now several cases where the complete kinetics of a catalytic reaction has been evaluated solely on the basis of DFT calculations of reaction barriers, reaction energies and the associated entropies[17, 18, 19, 20]. Figure 2 shows the comparison between calculated and measured rates for three different reactions and catalytic surfaces. Overall, the agreement between DFT-based kinetic models and experiment is surprisingly good, and they serve to illustrate the accuracy and value of current density functional theory.

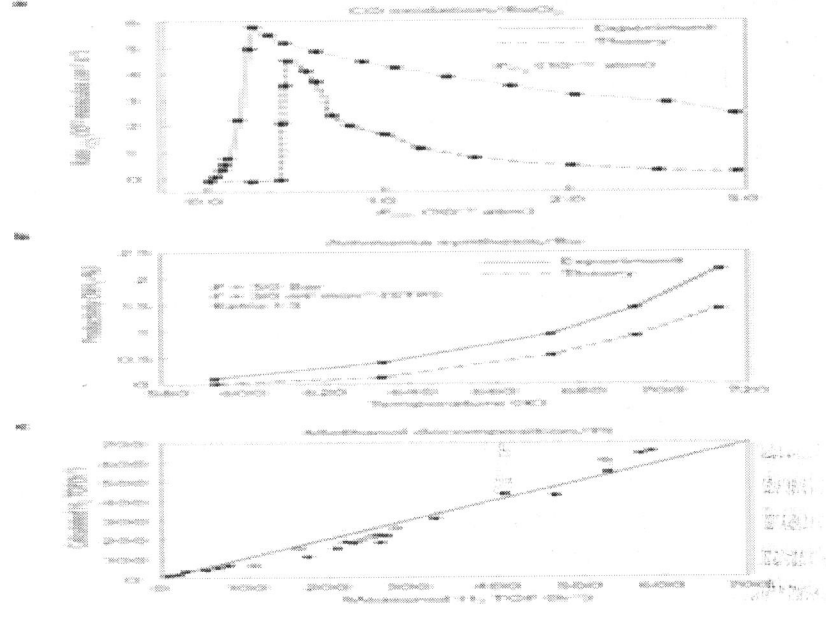

Figure 2: Comparison of experimental results for three different catalytic reactions with the results of kinetic models based on DFT calculations. a, CO oxidation activity over ruthenium oxide at low oxygen pressures. Adapted from ref. 18; © 2004 APS. b, Ammonia synthesis productivity over a ruthenium catalyst at industrial reaction conditions. Based on data from ref. 19. c, Methanol decomposition rate over a platinum catalyst. In each of these three cases the theoretical calculations and the experiments agree semi-quantitatively. Adapted from ref. 20; © 2006 Springer.

The agreement between theory and experiment is particularly noteworthy in two cases for supported metal catalysts (ruthenium and platinum in Fig. 2) — which are considerably more complex than a well-defined single crystal surface. Here, the theoretical treatment has assumed that the supported metallic nanoparticles can be viewed as crystalline particles with well-defined facets in addition to edges, corners, steps and kinks, and that these surface features can be treated as being independent of each other. Each surface structure will then contribute to the overall rate and the most active one will typically dominate. This is, for instance, the case for ammonia synthesis where step sites dominate[19]. Several experiments have shown real catalyst particles to have well-defined geometrical features[21, 22, 23, 24, 25]. The independence of the different types of surface sites on metal particles can be understood by noting that the electrostatic screening by the metallic, freely moving electrons introduces a 'nearsightedness'[26,27] such that a perturbation to the surface is only significant within a screening length — typically a few ångströms. For very small particles, where the electrons are no longer metallic, this picture breaks down — the exact size where this happens is still an open question.

The complete kinetic description of a given system is a quite demanding task. One cannot, at this moment, imagine screening a large number of systems using a procedure that requires such a description for each system considered. Rather, it is instructive to establish which properties at the atomic scale determine the macroscopic kinetics. Such an approach in terms of descriptors is outlined below.

The identification of descriptors is facilitated substantially by the observation that activation energies for elementary surface reactions are strongly correlated with adsorption energies. This is illustrated in Fig. 3 for the methanation reaction ($CO + 3H_2 \rightarrow CH_4 + H_2O$). First, it is established computationally that the activation barrier for CO dissociation is forbiddingly high on the most close-packed surface, whereas certain steps (and other similar geometries) have much lower barriers (by approximately 1 eV)[28, 29]. The active site on the catalyst surface is therefore identified as the steps or edges on the surface of the catalyst material.

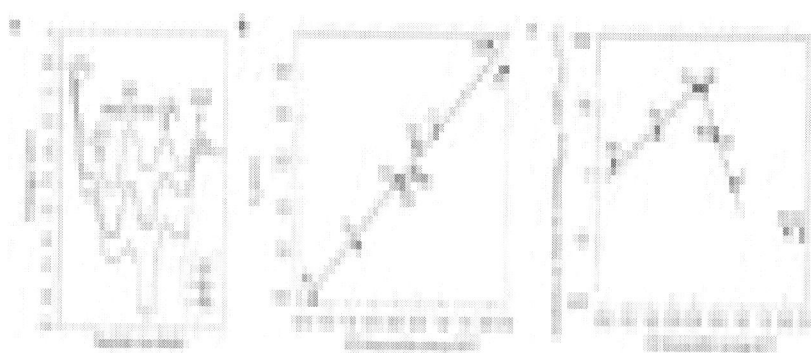

Figure 3: Identification of a descriptor for the methanation reaction ($CO + 3H_2 \rightarrow CH_4 + H_2O$). a, Calculated energy diagrams for CO methanation over nickel, ruthenium and rhenium. Only the highest of the activation barriers

for hydrogenation of C and O are included. b, Brønsted–Evans–Polanyi relation for CO dissociation over transition metal surfaces. The transition state potential energy, E_a, is linearly related to the CO dissociation energy. c, The corresponding measured volcano-relation for the methanation rate[33]. Parts b and c reprinted from ref.50; © 2006 Elsevier.

On comparing a series of different metal surfaces as catalysts for the methanation reaction (Fig. 3a) it is found that the barrier for CO activation, as well as the barriers for CH_4 and H_2O formation, are closely related to the stability of C and O on the surface. The more stable they are, the lower the barrier for CO dissociation will be, and the higher the barrier becomes for CH_4 and H_2O formation. In fact, all three activation energies are found to scale essentially linearly with the reaction energy in Brønsted–Evans–Polanyi (BEP)-type relationships (see Fig. 3b for CO dissociation)[28, 30, 31, 32]. Such correlations lead directly to a volcano relationship between the rate and the dissociative chemisorption energy, E_{diss}, of CO (ref. 33; see Fig. 3c). The reason is that in the limit of weak coupling (E_{diss} is only a little negative), the BEP relation gives that the barrier for dissociation of the reactants will be high and the rate low. For strong coupling (E_{diss} very negative) the activation energy of adsorption is small but now the barrier for forming the products will be large. An optimal interaction strength must exist between these two limits — this is known as the Sabatier principle[34]. Figure 3 shows that calculations can be used to quantify the interaction strength in such a way that experimental data for the methanation rates can be understood on this basis. E_{diss} is therefore a good descriptor for the catalytic activity of different catalysts for the methanation reaction, and we can identify its optimum value from Fig. 3.

In general there may be several descriptors, depending on the number of different important surface intermediates. The number of independent variables is limited strongly by the fact that it has been found that adsorption energies for a number of molecules scale with each other[35]. For the methanation reaction, for instance, the bond energy of adsorbed CH, CH_2 and CH_3 vary linearly with the bond energy of adsorbed C from one metal surface to the next, and the same is true for OH versus O adsorption energies.

Volcano relations between rates and adsorption energies have been widely identified in heterogeneous catalysis. For many years adsorption energies of intermediates were not readily available and various thermodynamic data, such as heats of oxide formation, were used as descriptors[36]. With the advent of sufficiently accurate DFT calculations this situation has completely changed, and descriptors of catalytic activity in terms of calculated adsorption energies have been identified for a number of systems[33, 37, 38].

The volcano-shaped relationships between total catalytic rates and adsorption energies may explain some of the good agreement between experiments and theory shown in Fig. 2. Close to the top of the volcano the rate depends only weakly on the absolute strength of the adsorption energies. For the methanation reaction, for instance, the window of values of E_{diss} around the maximum where the rate is within an order of magnitude of the maximum values is on the order of 0.5 eV. This means that for the best catalysts (close to the maximum of the volcano) errors of a few tenths of an eV may still give reasonable values for the rate. As this is the typical error of DFT calculations[15], they can give quite accurate rates at least close to the top of the volcano.

The Electronic Structure Factor

The variation in adsorption energy (and hence the catalytic activity) from one metal to the next is determined by the electronic structure of the surface. It turns out that for the transition metals the coupling between the adsorbate valence states and the metal d-states largely describe the variations[39, 40, 41, 42, 43, 44] The rule is that the higher in energy the d-states are relative to the highest occupied state — the Fermi energy — of the metal, the stronger the interaction with adsorbate states. The reason is that when the d-states are close to the Fermi energy, antibonding states can be shifted well above it and become empty (or bonding states can be shifted below it and become occupied). This increases the bond strength. Figure 4a establishes how variations in adsorption energy from one metal to the next are correlated with shifts in the energy of the d-states. Figure 4b,c shows a more subtle effect: The electronic structure of a platinum surface can be engineered by inserting another metal (nickel, cobalt, iron and so on) in the second layer and this directly affects the oxygen and hydrogen adsorption energies. It shows how changing the metal ligands of the surface platinum atoms can change its chemical properties substantially.

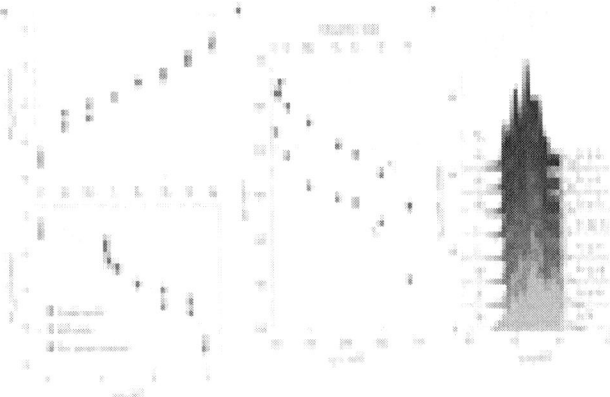

Figure 4: The d-band model — understanding the electronic origin of variations in surface chemistry. a, Variations in the O adsorption energy, $E_{ads}(O)$, on the most close-packed surface of the 4d transition metal series. The results of full DFT calculations are compared with those from a simple Newns–Anderson model[98] and to experiments (polycrystalline surfaces)[9]. Below, the same data are plotted as a function of the average energy of the d-electrons (the d-band centre with respect to the Fermi level), $\varepsilon_d - \varepsilon_F$, on the transition metal. Adapted from ref. 16; © 2000 Elsevier. b, Calculated changes in the dissociative adsorption energy of H_2 and O_2, E, versus the average energy of the projected density of states for the surface platinum d-states. c, Local projected densities of states, $n(\ )$, for a series of Pt(111) surfaces, where the second layer has been replaced by a layer of a 3d transition metals are shown. N_d is the number of d electron on the surface Pt atoms, which is hardly affected by the subsurface atoms. Parts b and c adapted from ref. 99; © 2004 Elsevier.

Catalyst Design

The first examples of where ideas generated from electronic structure calculations were exploited in the search for new solid catalysts include: the modification of the stability of Ni catalysts for steam reforming by the addition of gold[45]; the mixing of cobalt and molybdenum in ammonia synthesis catalysts[46]; new mixed transition metal sulfides for hydro-desulfurization[47]; new CO-tolerant alloys for fuel-cell anodes[48]; and near-surface alloys for hydrogen activation[49].

The first example of extensive computational screening of surface structures for new catalysts was for the methanation reaction[50]; this reaction is used extensively in industry to remove trace amounts of CO from hydrogen streams produced by steam-reforming of hydrocarbons[51].

The approach taken was as follows. First the CO dissociation energy, E_{diss}, was identified as a descriptor of catalytic activity as described above, and indicated in Fig. 3. The optimal value was identified by comparison to experimental data for the elemental metals, see Fig. 3c. Then a series of binary alloys (with concentration varying in steps of 25%) were formed from metals (Ni, Pd, Pt, Co, Rh, Ir, Fe, Ru and Re) chosen so that they should be reasonably stable at methanation conditions. For each alloy the catalytic performance descriptor $|E_{diss} - E_{diss}(\text{optimal})|$ was then calculated using a simplified interpolation model. A total of 117 different systems were studied.

In the case of the methanation reaction, there are already elemental metals, ruthenium and cobalt, close to the top of the volcano, see Fig. 3c. These metals are, however, not used industrially because they are quite costly. Instead the cheaper but also inferior catalyst material Ni is used. The cost of the raw materials is therefore an important parameter, and in Fig. 5a all the alloys and elemental metals included in the study are shown in a cost versus catalytic performance plot. NiFe alloys stand out in this plot as having a high catalytic activity as well as a low price. They were therefore chosen for a more detailed theoretical and experimental study. This involved a full DFT calculation of the energetics to make sure that the simple interpolation model was correct. It also involved a series of computational tests of stability of the alloy towards segregation. The result of the experimental test is included in Fig. 5b. A series of catalysts supported on MgAl spinel were prepared and their methanation activities were measured. It can be seen that the NiFe alloys are indeed more active than both pure nickel and iron, as predicted. Subsequently, the concept was converted into a technical catalyst at Haldor Topsøe[52].

Towards the Computational Design of Solid Catalysts

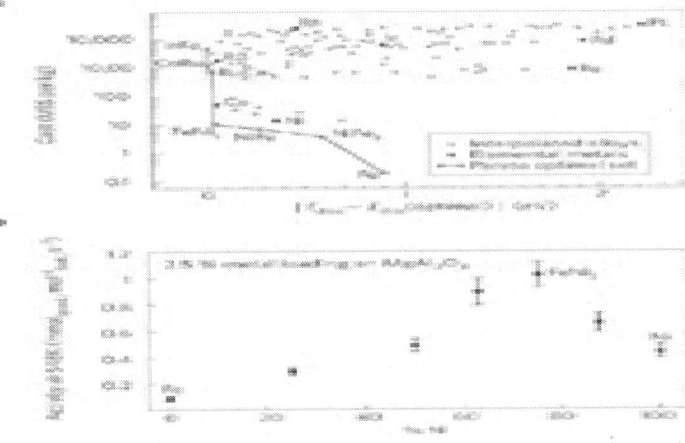

Figure 5: Computational screening for methanation catalysts. a, A price versus catalytic-performance plot for methanation over a range of elemental metals and alloys. The closer the descriptor E_{diss} (the CO dissociation energy) is to the optimum value (the smaller the value of $|E_{diss}-E_{diss}(\text{optimal})|$ is) the better the predicted catalytic activity. The Pareto optimal set of solutions is connected by the solid line, which defines the best compromise between price and catalytic performance for the set of systems investigated theoretically. b, Experimental confirmation that NiFe alloys are more active than pure nickel. The error bars indicate the estimated standard deviation of the measured rate of 10%. Adapted from ref. 50; © 2006 Elsevier.

An Example from Electrocatalysis

Electro-catalysis design is currently attracting much attention mainly for energy-conversion purposes. Many future energy transformation processes rely on electro-catalysis. One important example is the evolution of hydrogen in electrolysis and the reverse process where hydrogen is used as a fuel in a fuel cell. In acidic solutions platinum is the preferred catalyst material for both processes. As a hydrogen electrode it is stable and effective, but it is scarce and expensive, and extensive research efforts are directed towards replacing it — or at least reducing the amount needed.

Compared with catalysts for gas-phase reactions, the description of electro-catalysts has additional challenges due to the liquid phase in direct contact with the catalysts surface and due to charging of the surface[53, 54, 55, 56, 57, 58]. Another very important constraint is the corrosive environment that the catalyst is often exposed to in the electrolyte. Many of the non-precious catalyst materials important in conventional heterogeneous catalysis, for example, iron, cobalt or nickel, will quickly dissolve in acids.

The hydrogen-evolution reaction, where protons and electrons recombine to form molecular H_2, is one of the simplest electrochemical reactions, but still no good alternative to the platinum catalyst has been found. The adsorption free-energy of hydrogen, G_{H*}, is a good descriptor for hydrogen evolution[59, 60,61]. This makes sense because no matter what the reaction path is, adsorbed hydrogen is one of the intermediates. If H binds too weakly to the surface, H^+ cannot adsorb from the dissolved phase and if it binds too strongly, it will have difficulty leaving the surface for the gas phase. One would expect the optimal rate when hydrogen at the surface is as stable as gas-phase hydrogen — which by definition has the same free energy as solvated protons and electrons at zero potential relative to the normal hydrogen electrode (see Fig. 6a). Plotting the exchange current density versus the binding of hydrogen obtained by DFT indeed shows a volcano with an optimum around zero free energy of adsorption[62] (see Fig. 6b).

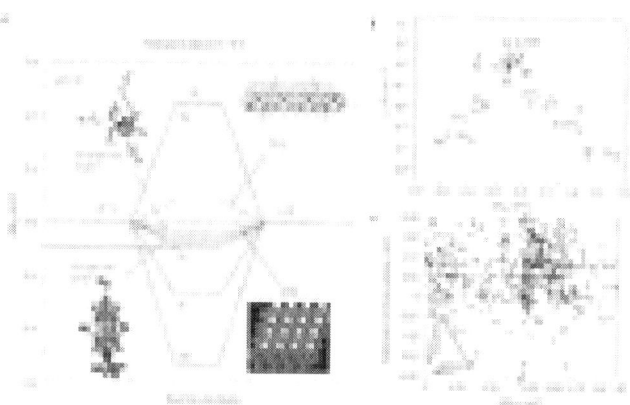

Figure 6: Screening for hydrogen-evolution catalysts. a, The free energy diagram of hydrogen evolution at zero potential and zero pH for gold, platinum, nickel, molybdenum and PtBi close-packed surfaces, the MoS_2 edge, and the active centres in hydrogenase and nitrogenase. The closer the binding free energy of the intermediate — where H

atoms are bound to the catalyst — is to zero the higher the activity. Adapted from ref. 63; © 2005 ACS. b, The experimental exchange current, i_0, is plotted as a function of the calculated standard free energy of adsorption of hydrogen, G_{H^*}. Experimental data from many different experiments are included, which accounts for the scatter. In one particular set of experiments (marked in blue) platinum and a PtBi surface alloy are compared. Adapted from ref. 71; © 2007 AAAS. c, The stability of different surface alloys is plotted as a function of the binding free energy of hydrogen. In the lower left quadrant are the stable and active surface alloys and the points that limit the set from lower left is the Pareto optimal set. Adapted from ref. 64; © 2006 NPG.

A computational search for high activity can then be carried out by calculating G_{H^*}. As stability of the catalyst is a major issue, the calculation of descriptors for stability is as important as for activity. A range of surface alloys (alloys only in the first layer) with the optimal combined stability and activity can then be indentified[63, 64] (see Fig. 6c). One interesting candidate is a surface alloy of platinum and bismuth. Supported on pure platinum, adsorbed bismuth is known to poison hydrogen evolution[65], however, when the surface is annealed, a PtBi surface alloy is formed showing a measured activity slightly higher than that of a reference sample of pure platinum[64].

Another strategy for identifying materials that could have promising features as hydrogen-evolution catalysts is by taking inspiration from biology. Hydrogenases[66] and nitrogenases[67] are known to be good catalysts for hydrogen evolution. The descriptor approach also applies to the active centres of enzymes[63, 68] (see Fig. 6a). Both hydrogenases and nitrogenases have catalytic sites containing sulfur atoms bridged between metal atoms. In looking for inorganic analogues to the active centre in the enzymes it was noted that the same arrangement for sulfur is found at the edge of MoS_2 slabs or nanoparticles. These structures are well-known as hydro-desulfurization catalysts used in removing sulfur-containing molecules from oil products[69, 70]. The MoS_2 particles supported on carbon and gold have been tested showing that hydrogen evolution is indeed possible on MoS_2 (refs 63, 71; see Fig. 6b).

Addressing Selectivity

Often selectivity towards specific products is of key interest. Selective processes do not only offer cleaner chemistry and better environmental protection, but also allow for improving the use of resources thus leading to more economic production[72].

As selectivity is related to favouring specific reaction pathways among several competing pathways, a prerequisite for the theoretical treatment of selectivity is the accurate treatment of the activity of single reaction pathways. This treatment has to be accomplished at least with sufficient accuracy to address relative changes in the energy barriers between competing pathways.

Ethylene Oxide Synthesis. Ethylene oxide (EO) is an important chemical with an annual global production of the order of 10 million tons[73]. It is primarily used in organic synthesis reactions. All large-scale production of ethylene oxide is today done by direct partial oxidation of ethylene over a silver catalyst[74]. The selectivity of a typical catalytic EO process is 65% to 80% depending on whether the oxidant is air or pure O_2 (ref. 73). The side product is mainly the full combustion product, CO_2. As the primary expense in the process is the ethylene cost, high selectivity towards EO is important in improving cost-efficiency and minimizing CO_2 emissions.

High-resolution electron energy loss spectroscopy (HREELS) experiments and DFT calculations have shown that an oxametallacycle[75] species is a key intermediate in the ethylene oxide formation over Ag(111) (ref. 76). This has enabled the construction of a detailed DFT-based kinetic model that agrees well with ethylene oxidation rate experiments over Ag (ref. 77). Two competitive transition states lead to ethylene oxide and acetaldehyde, respectively, see Fig. 7a,b. The acetaldehyde eventually goes to full combustion, whereas EO directly desorbs and is unlikely to react further. The difference in energy between these two transition states thus becomes a good descriptor for the selectivity of an EO catalyst, and catalysts, which favour the transition state going towards EO, can be sought computationally[78]. InFig. 7c the difference in the two transition state energies relative to the difference over silver is shown for a few bimetallic Ag catalysts. It is observed that some presence of copper atoms in the silver surface should yield particularly high selectivity towards EO. The calculations were subsequently verified through the synthesis and testing of a number of Cu/Ag-containing surface alloys. The results are shown in Fig. 7d. It is observed that as the bulk contents of copper increases slightly, the selectivity increases by almost 50% compared with a pure silver reference catalyst[78].

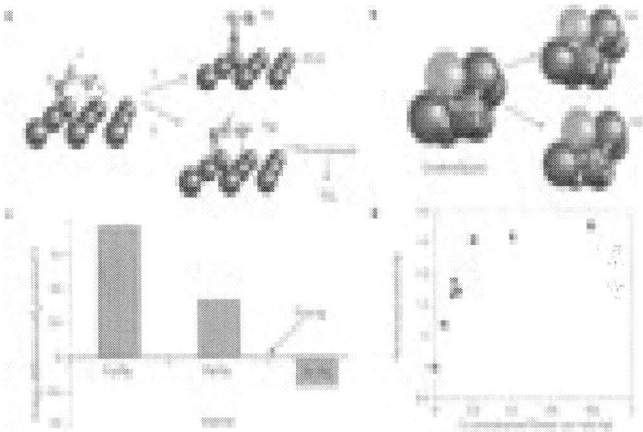

Figure 7: Computational design of ethylene oxide (EO) synthesis catalysts with improved selectivity. a, Competing oxametallacycle pathways. Activation barriers for the two pathways are calculated to be similar over silver. b, Structure of a bimetallic model catalysts. TS is transition state. c, The selectivity descriptor, $E_A = (E_{TS2}(\text{alloy}) - E_{TS1}(\text{alloy})) - (E_{TS2}(Ag) - E_{TS1}(Ag))$, shown for a number of catalyst compositions. Higher descriptor values means that the bimetallic should be more selective than pure silver. d, The measured selectivity relative to pure silver as a function of bulk copper content. All parts adapted from ref. 78; © 2004 Elsevier.

Preferential Oxidation of CO in Hydrogen. Preferential oxidation of CO in hydrogen (PROX) currently attracts significant attention as an alternative to methanation for removing CO from hydrogen, in particular for fuel-cell applications. The PROX reaction is carried out in a large excess of hydrogen, and the reaction can for example be written as:

$$CO + \tfrac{1}{2}O_2 + H_2 \rightarrow CO_2 + H_2$$

specifying that hydrogen is not consumed in the process. It is very difficult in practice to avoid some hydrogen being oxidized into water. A highly selective catalyst is thus desirable to reduce the amount of CO to an adequate level without combusting too much of the valuable hydrogen. This is of particular importance for hydrogen-consuming applications such as hydrogen proton exchange membrane fuel cells, where even a few tens of ppm CO will poison currently used Pt-based electrocatalysts[79].

On the basis of DFT studies, core–shell nanoparticles have been proposed as candidates for new catalytic properties different from pure metal surfaces, surface alloys and near-surface alloys[80]. Detailed computational studies of platinum-covered ruthenium, iridium, rhodium, palladium, gold and platinum were carried out. These studies suggested that Pt-covered ruthenium, so-called Ru@Pt, could present unique features compared with the other core–shell structures and the pure platinum nanoparticles, as the binding of CO molecules were significantly weakened on the Ru@Pt. The effect of the ruthenium underneath the platinum surface on the CO adsorption is the same electronic effect discussed in connection with Fig. 4: the platinum d-states are shifted up in energy due to the ruthenium atoms, and this ligand effect[81] changes the CO bond strength. Experiments have shown that the reaction temperature is significantly lower for PROX over Ru@Pt particles than PtRu alloy, as predicted from calculations. Experiments also show that 70% of the CO is already oxidized to CO_2 at 30°C over the Ru@Pt (ref. 80).

Selective hydrogenation of acetylene. Large-scale production of ethylene is primarily carried out by steam-cracking of saturated hydrocarbons[73] which leads to impurities in the form of acetylene in the ethylene product slate. Much of the ethylene is used in processes where acetylene is undesirable. One process where the acetylene is particularly undesired is the important polymerization of ethylene into polyethylene. The acetylene concentration in the ethylene feed can be reduced by selective hydrogenation to ethylene. A high selectivity is necessary to get the acetylene reduced to the desired low levels of a few ppm without hydrogenating ethylene to ethane. The most commonly used catalyst in industry is a silver-modified palladium catalyst.

Density functional theory calculations for a number of transition-metal surfaces show that acetylene and ethylene adsorption energies scale with methyl adsorption energies[82] (Fig. 8a). The slope of the scaling relations in the reactive surface regime is four for C_2H_2 and two for C_2H_4. This can be viewed as a manifestation of bond-order conservation for the surface-bonded carbon atoms[35]. The scaling relations are thus related to bond-order conservation models[83]. A good acetylene hydrogenation catalyst should present a high stability of adsorbed acetylene and a low stability of ethylene. Strong acetylene binding leads to high acetylene removal rate, whereas weak ethylene adsorption leads to ethylene being desorbed instead of further hydrogenation, and therefore high selectivity. This, together with the scaling relations, leads to a window of simultaneously active and selective catalysts as expressed by using the methyl binding energy as a descriptor (see Fig. 8a).

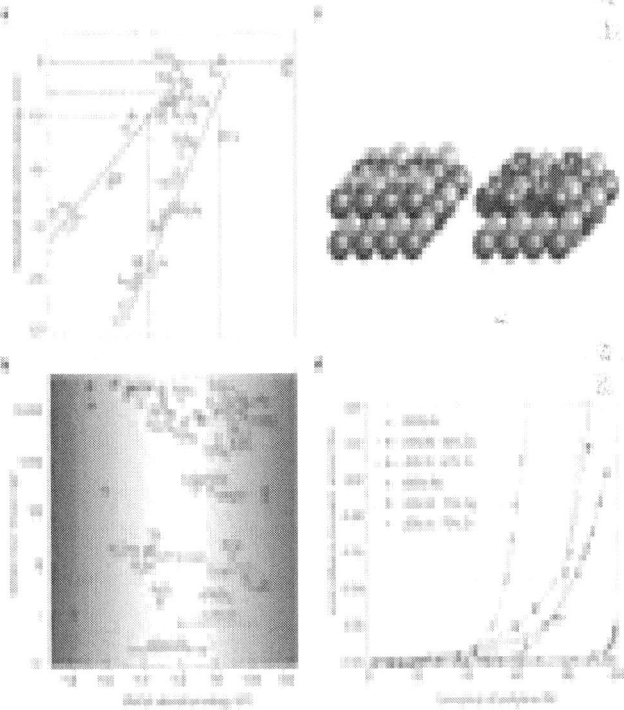

Figure 8: Identifying catalysts for selective acetylene hydrogenation. a, The calculated binding of acetylene (C_2H_2) and ethylene (C_2H_4) as a function of methyl (CH_3) adsorption over a number of metals and alloys. The solid lines for acetylene (red) and ethylene (blue) identify linear scaling relations. The dotted blue line defines the maximal methyl binding for which the ethylene

is predicted to desorb more easily than hydrogenating to ethane. The dotted red line identifies the minimal methyl binding necessary to obtain a turnover frequency on the order of 1 s^{-1}. Together the lines define a window of methyl binding in which catalysts that are simultaneously active and selective are predicted to lie. b, Constituent cost of 70 binary intermetallic compounds as a function of their calculated methyl adsorption energies. c, The adsorption of acetylene (left) and ethylene (right) on a (bcc-B$_2$ (110)) structural model of the NiZn catalyst (nickel atoms: blue; zinc atoms: grey). d, The measured concentration of ethane at the reactor outlet as a function of acetylene conversion for seven different catalysts. Zero ethane at high conversion is desirable. The experiments verify the theoretical prediction that NiZn catalysts could exhibit selectivity comparable to that of PdAg. All parts adapted from ref. 82; © 2008 AAAS.

Screening of approximately 70 different alloy surfaces for their methyl binding energies yielded the results shown in Fig. 8b where the constituent cost is plotted versus methyl adsorption energy. A number of alloys fall in the window of interest, including several PdAg alloys, as expected. Also identified are the alloys made from PdGa, PdPb and PdAu, which have recently been shown experimentally to exhibit a good activity and selectivity[84, 85, 86, 87]. The alloys CoGa, NiGa, FeZn and NiZn stand out as particularly interesting, because they seem to be active, selective and inexpensive. An analysis of the stability of the different alloys shows that the NiZn alloys are particularly stable, and the NiZn compounds were therefore chosen for further study. In Fig. 8c, the adsorption structures of acetylene and ethylene on the Ni–Zn alloy are shown. The adsorbates are bonded to the nickel sites, which show that the change in adsorption properties is not a result of bonding to the zinc. Instead, the zinc atoms change the electronic properties of the nickel atoms.

A series of NiZn alloy catalysts on MgAl$_2$O$_4$ spinel supports were synthesized and tested for their selectivity in the hydrogenation of acetylene in a gas mixture of ethylene, acetylene and hydrogen. The ethane production as a function of acetylene conversion is shown in Fig. 8d. A highly selective catalyst will have very low ethane production, even at high conversion, where the amount of acetylene in the reactants is small. Different NiZn catalysts were compared with a model PdAg catalyst. Pure Pd has a reasonably good selectivity, but the PdAg alloy shows a very high selectivity even at high conversions. Nickel is considerably worse than palladium, but as expected from

Fig. 8b, adding increasing amounts of zinc increases the selectivity substantially. The NiZn catalyst with the highest zinc content had selectivity comparable to the best PdAg catalyst that was tested.

Outlook

The fact that it has been possible to tailor surfaces with improved catalytic properties from theoretical insights and DFT calculations provides some hope that this may develop into a more generally versatile design strategy. There are, however, a number of challenges ahead.

First, it should be realized that finding leads for new catalysts is only one step towards a new technical catalyst. High catalytic activity or selectivity and low constituent cost can be necessary requirements for a new catalyst, but long-term stability, lack of side-products, resistance to poisons, susceptibility to promoters and cost of production are equally important factors. To some extent these factors may also be simulated, but in the end, experimental studies under realistic conditions will always be central to creating technical catalysts.

An important extension of the notion of DFT-based catalyst design is the use of DFT calculations in reactor design. The first steps in this direction were taken for the ammonia synthesis process in which it proved possible to link the atomic-scale insight obtained by DFT calculations directly with the industrial chemical engineering practice as illustrated in Fig. 9. In an industrial ammonia synthesis reactor there are several catalyst beds with cooling stages in between as illustrated in Fig. 9a. The cooling stages are introduced so as to operate as close to the maximum rate line (red in Fig. 9a) as possible. The important notion is that the position of the maximum for the volcano curve (Fig. 9b) is a strong function of the operating conditions. At low ammonia concentrations (reactor inlet), Fe is the preferred elemental catalyst, whereas at high ammonia concentrations (reactor outlet), Ru is the preferred elemental catalyst. The optimal catalyst curves (Fig. 9c) express the properties of the optimal catalyst at given reaction conditions plotted with the operating line. Thus, this illustrates the value(s) of the activity descriptor(s) at the maximum of the volcano curve at the given reaction conditions. The key concept is that the structure and composition of the optimal catalyst is a function of the reaction conditions, and as these vary throughout industrial reactors, it is desirable to perform the computational screening as a function of all possible reaction conditions. This might also, in a longer perspective, be a way to identify radically new catalysts rather than simply improving the performance of known catalysts.

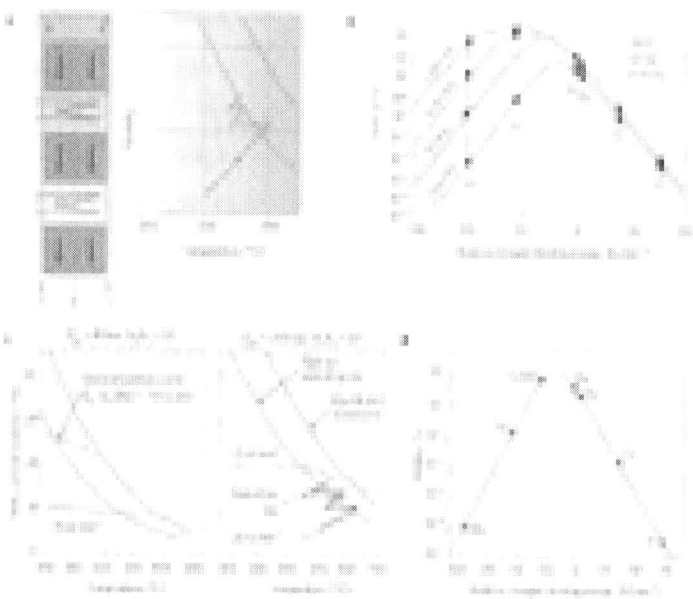

Figure 9: Linking DFT calculations with industrial reactor design and catalyst selection. a, Schematic illustration of an industrial ammonia synthesis reactor with three adiabatic catalyst beds and two cooling stages. The diagram shows the equilibrium line, the optimal operating line, which can also be called maximum rate line (red), and the operating line for the reactor configuration shown. The closer the operating line approaches the optimal operating line, the lower the catalyst volume required. b, Volcano curves for the turnover frequency, TOF, calculated based on micro-kinetic modelling using parameters calculated by DFT. c, An optimal catalyst curve expresses the reaction conditions under which a transition metal surface with a given nitrogen binding energy is the theoretically optimal catalyst. The optimal catalyst curves make it possible to identify desirable catalysts for relevant reaction conditions. E_{N^*} is the metal–nitrogen binding energy (negative values signify exothermic adsorption). d, The interpolation concept illustrating

that the binding energy for a CoMo catalyst is intermediate between that of the elemental catalysts Co and Mo (ref. 46). With this concept, or other alloy models, it is possible to identify suitable catalyst leads to be used with the optimal catalyst curves, that is, to design catalysts for specific rector design and process conditions. Figure adapted from ref.100; © 2002 Elsevier.

The examples discussed in this review all refer to catalysts where the active site consists of a transition metal (alloy) surface. It is implicitly assumed that the surface of the supported nanoparticles can be modelled by extended surfaces, perhaps with defects. Much more work is required to find out when this assumption breaks down and how one then systematically includes support effects directly in the simulations. Going beyond transition metal catalysts may also provide a considerable challenge from a theoretical point of view. From detailed comparisons between theory and experiment we know that DFT works quite well for these systems, but we also know that it may work less well for other classes of catalysts including for example some strongly correlated oxides[88, 89, 90, 91]. There are also a number of situations where it is essential to include van der Waals interactions[92]. Recent developments suggest that this may become a possibility in the near future[93, 94]. Additional challenges are related to finding methods to determine the ground-state structures of unknown materials[95, 96, 97]. We also need to describe the interaction of more complicated molecules with all these possible surface structures, and additional complications arise in describing electrocatalytic and photocatalytic processes.

Although experimental methods usually tend to become more expensive with time, computational methods will become cheaper as computers become faster. In combination with new developments in electronic structure theory and computational methods, this suggests that computational approaches for the discovery and development of catalysts hold great promise for the future.

ACKNOWLEDGEMENTS

The Center for Atomic-scale Materials Design is funded by the Lundbeck Foundation. J.K.N. acknowledges support from Ib Henriksens Fond.

REFERENCES

1. Thomas, J. M. & Thomas, W.-J. *Principle and Practice of Heterogeneous Catalysis* (VCH, 1997).
2. Ceder, G. et al. Identification of cathode materials for lithium batteries guided by first-principles calculations. *Nature* 392, 694–696 (1998).
3. Alapati, S. V., Johnson, J. K. & Sholl, D. S. Using first principles calculations to identify new destabilized metal hydride reactions for reversible hydrogen storage. *Phys. Chem. Chem. Phys.* 9, 1438–1452 (2007).
4. Piquini, P., Graf, P. A. & Zunger, A. Band-gap design of quaternary (In, Ga)(As, Sb) semiconductors via the inverse-band-structure approach. *Phys. Rev. Lett.* 100, 186403 (2008).
5. Wondimagegn, T., Wang, D., Razavi, A. & Ziegler, T. Computational design of C-2-symmetric metallocene-based catalysts for the synthesis of high molecular weight polymers from ethylene/propylene copolymerization. *Organometallics* 27, 6434–6439 (2008).
6. Conley, B. L. et al. Design and study of homogeneous catalysts for the selective, low temperature oxidation of hydrocarbons. *J. Mol. Cat. A-Chem.* 251, 8–23 (2006).
7. Nilsson, A., Pettersson, L. G. M. & Nørskov, J. K. (eds) *Chemical Bonding at Surfaces and Interfaces* (Elsevier, 2008).
8. Ertl, G. Reactions at surfaces: from atoms to complexity. *Angew. Chem. Int. Ed.* 47, 3524–3535 (2008).
9. Somorjai, G. A. *Introduction to Surface Chemistry and Catalysis* (Wiley, 1994).
10. Yeo, Y. Y., Vattuone, L. & King, D. A. Calorimetric heats for CO and oxygen adsorption and for the catalytic CO oxidation reaction on Pt{111}. *J. Chem. Phys.* 106, 392–401 (1997).
11. Goodman, D. W., Kelley, R. D., Madey, T. E. & Yates, J. T. Kinetics of the hydrogenation of CO over a single crystal nickel-catalyst. *J. Catal.* 63, 226–234 (1980).
12. Lytken, O. et al. Energetics of cyclohexene adsorption and reaction on Pt(111) by low-temperature microcalorimetry. *J. Am. Chem. Soc.* 130, 10247–10257 (2008).

13. Kohn, W. & Sham, L. J. Self-consistent equations including exchange and correlation effects. *Phys. Rev. A* 140, 1133–1138 (1965).
14. Perdew, J. P., Burke, K. & Ernzerhof, M. Generalized gradient approximation made simple. *Phys. Rev. Lett.* 77, 3865–3868 (1996).
15. Hammer, B., Hansen, L. B. & Nørskov, J. K. Improved adsorption energetics within density functional theory using revised PBE functionals. *Phys. Rev. B* 59, 7413–7421 (1999).
16. Hammer, B. & Nørskov, J. K. Theoretical surface science and catalysis — calculations and concepts. *Adv. Catal.* 45, 71–129 (2000).
17. Hansen, E. W. & Neurock, M. First-principles-based Monte Carlo simulation of ethylene hydrogenation kinetics on Pd. *J. Catal.* 196, 241–252 (2000).
18. Reuter, K., Frenkel, D. & Scheffler, M. The steady state of heterogeneous catalysis, studied by first-principles statistical mechanics. *Phys. Rev. Lett.* 93, 116105 (2004).
19. Honkala, K. et al. Ammonia synthesis from first-principles calculations. *Science* 307, 555–558 (2005).
20. Kandoi, S. et al. Prediction of experimental methanol decomposition rates on platinum from first principles. *Top. Catal.* 37, 17–28 (2006).
21. Hansen, K. H. et al. Palladium nanocrystals on Al_2O_3: Structure and adhesion energy. *Phys. Rev. Lett.* 83, 4120–4123 (1999).
22. Hansen, T. W. et al. Atomic-resolution *in situ* transmission electron microscopy of a promoter of a heterogeneous catalyst. *Science* 294, 1508–1510 (2001).
23. Hansen, P. L. et al. Atom-resolved imaging of dynamic shape changes in supported copper nanocrystals. *Science* 295, 2053–2055 (2002).
24. Hofmann, S. et al. In situ observations of catalyst dynamics during surface-bound carbon nanotube nucleation. *Nano Lett.* 7, 602–608 (2007).
25. Gontard, L. C. et al. Aberration-corrected imaging of active sites on industrial catalyst nanops. *Angew. Chem. Int. Ed.* 46, 3683–3685 (2007).

26. Kohn, W. Density functional and density matrix method scaling linearly with the number of atoms. *Phys. Rev. Lett.* 76, 3168–3171 (1996).
27. Prodan, E. & Kohn, W. Nearsightedness of electronic matter. *Proc. Natl Acad. Sci. USA* 102, 11635–11638 (2005).
28. Nørskov, J. K. et al. Universality in heterogeneous catalysis. *J. Catal.* 209, 275–278 (2002).
29. Ciobica, I. M. & van Santen, R. A. Carbon monoxide dissociation on planar and stepped Ru(0001) surfaces. *J. Phys. Chem. B* 107, 3808–3812 (2003).
30. Pallassana, V. & Neurock, M. Electronic factors governing ethylene hydrogenation and dehydrogenation activity of pseudomorphic Pd-ML/Re(0001), Pd-ML/Ru(0001), Pd(111), and Pd-ML/Au(111) surfaces. *J. Catal.* 191, 301–317 (2000).
31. Alcalá, R., Mavrikakis, M. & Dumesic, J. A. DFT studies for cleavage of C-C and C-O bonds in surface species derived from ethanol on Pt(111). *J. Catal.* 218, 178–190 (2003).
32. Michaelides, A. et al. Identification of general linear relationships between activation energies and enthalpy changes for dissociation reactions at surfaces. *J. Am. Chem. Soc.* 125, 3704–3705 (2003).
33. Bligaard, T. et al. The Brønsted-Evans-Polanyi relation and the volcano curve in heterogeneous catalysis. *J. Catal.* 224, 206–217 (2004).
34. Sabatier, P. Hydrogénations et déshydrogénations par catalyse. *Ber. Deutsch. Chem. Gesellshaft* 44, 1984–2001 (1911).
35. Abild-Pedersen, F. et al. Scaling properties of adsorption energies for hydrogen-containing molecules on transition-metal surfaces. *Phys. Rev. Lett.* 99, 016105 (2007).
36. Boudart, M. in *Handbook of Heterogeneous Catalysis* (eds Ertl, G., Knözinger, H. & Weitkamp, J.) 1 (Wiley-VCH, Weinheim, 1997).
37. Falsig, H. et al. Trends in the catalytic CO oxidation activity of nanops. *Angew. Chem. Int. Ed.* 47, 4835–4835 (2008).
38. Cheng, J. & Hu, P. Utilization of the three-dimensional volcano surface to understand the chemistry of multiphase systems in heterogeneous catalysis. *J. Am. Chem. Soc.* 130, 10868–10869 (2008).

39. Holloway, S., Lundqvist, B. I. & Nørskov, J. K. in *Proc. 8th Conference on Catalysis, Berlin* vol.IV, p.85 (Verlag Chemie, 1984).
40. Hammer, B. & Nørskov, J. K. Why gold is the noblest of all the metals. *Nature* 376, 238–240 (1995).
41. Mavrikakis, M., Hammer, B. & Nørskov, J. K. Effect of strain on the reactivity of metal surfaces. *Phys. Rev. Lett.* 81, 2819–2822 (1998).
42. Roudgar, A. & Gross, A. Local reactivity of metal overlayers: Density functional theory calculations of Pd on Au. *Phys. Rev. B* 67, 33409 (2003).
43. Gajdos, M., Eichler, A. & Hafner, J. CO adsorption on close-packed transition and noble metal surfaces: trends from ab initio calculations. *J. Phys. Condens. Matter* 16, 1141–1164 (2004).
44. Nilsson, A. *et al.* The electronic structure effect in heterogeneous catalysis. *Catal. Lett.* 100 111–114 (2005).
45. Besenbacher, F. *et al.* Design of a surface alloy catalyst for steam reforming. *Science* 279, 1913–1915 (1998).
46. Jacobsen, C. J. H. *et al.* Catalyst design by interpolation in the periodic table: Bimetallic ammonia synthesis catalysts. *J. Am. Chem. Soc.* 123, 8404–8405 (2001).
47. Toulhoat, H. & Raybaud, P. Kinetic interpretation of catalytic activity patterns based on theoretical chemical descriptors. *J. Catal.* 216, 63–72 (2003).
48. Strasser, P. *et al.* High throughput experimental and theoretical predictive screening of materials. A comparative study of search strategies for new fuel cell anode catalysts. *Phys. Chem. B* 107, 11013–11021 (2003).
49. Greely, J. & Mavrikakis, M. Alloy catalysts designed from first principles. *Nature Mater.* 3, 810–815 (2004).
50. Andersson, M. P. *et al.* Toward computational screening in heterogeneous catalysis: Pareto-optimal methanation catalysts. *J. Catal.* 239, 501–506 (2006).
51. Sabatier, P. & Senderens, J. B. New methane synthesis. *Compte Rendu Acad. Sci. Paris* 134, 514–516 (1902).
52. Sehested, J. *et al.* Discovery of technical methanation catalysts based on computational screening. *Top. Catal.* 45, 9–13 (2007).

53. Jinnouchi, R. & Anderson, A. B. Aqueous and surface redox potentials from self-consistently determined Gibbs energies. *J. Phys. Chem. C* 112, 8747–8750 (2008).
54. Rossmeisl, J., Skúlason, E., Björketun, M. E., Tripkovic, V. & Nørskov, J. K. Modeling the electrified solid-liquid interface. *Chem. Phys. Lett.* 466, 68–71 (2008). | | |
55. Shubina, T. E. & Koper, M. T. M. Co-adsorption of water and hydroxyl on a Pt_2Ru surface.*Electrochem. Commun.* 8, 703–706 (2006).
56. Roudgar, A. & Gross, A. Water bilayer on the Pd/Au(111) overlayer system: Coadsorption and electric field effects. *Chem. Phys. Lett.* 409, 157–162 (2005).
57. Sugino, O. *et al.* First-principles molecular dynamics simulation of biased electrode/solution interface. *Surf. Sci.* 601, 5237–5240 (2007).
58. Filhol, J. S. & Neurock, M. Elucidation of the electrochemical activation of water over Pd by first principles. *Angew. Chem. Int. Ed.* 45, 402–406 (2006).
59. Conway, B. E. & Bockris, J. O. M. Electrolytic hydrogen evolution kinetics and its relation to the electronic and adsorptive properties of the metal. *J. Chem. Phys.* 26, 532–541 (1957).
60. Parsons, R. The rate of electrolytic hydrogen evolution and the heat of adsorption of hydrogen.*Trans. Faraday Soc.* 54, 1053–1063 (1958).
61. Trasatti, S. Work function, electronegativity, and electrochemical behavior of metals.3. Electrolytic hydrogen evolution in acid solutions. *J. Electroanal. Chem.* 39, 163–184 (1972).
62. Nørskov, J. K. *et al.* Trends in the exchange current for hydrogen evolution. *J. Electrochem. Soc.*152 J23–J26 (2005).
63. Hinnemann, B. *et al.* Biomimetic hydrogen evolution. *J. Am. Chem. Soc.* 127, 5308–5309 (2005).
64. Greeley, J. *et al.* Computational high-throughput screening of electrocatalytic materials for hydrogen evolution. *Nature Mater.* 5, 909–913 (2006).
65. Gómez, R., Feliu, J. M. & Aldaz, A. Effects of irreversibly adsorbed bismuth on hydrogen adsorption and evolution on Pt(111). *Electrochim. Acta* 42, 1675–1683 (1997).

66. Evans, D. J. & Pickett, C. J. Chemistry and the hydrogenases. *Chem. Soc. Rev.* 32, 268–275 (2003).
67. Rees, D. C. & Howard, J. B. The interface between the biological and inorganic worlds: Iron-sulfur metalloclusters. *Science* 300, 929–931 (2003).
68. Siegbahn, P. E. M., Tye, J. W. & Hall, M. B. Computational studies of [NiFe] and [FeFe] hydrogenases. *Chem Rev.* 107, 4414–4435 (2007).
69. Topsøe, H., Clausen, B. S. & Massoth, F. E. in *Catalysis, Science and Technology* Vol. 11 (eds Anderson, J. R. & Boudart, M.) 1–310 (Springer, 1996).
70. Helveg, S. et al. Atomic-scale structure of single-layer MoS_2 nanoclusters. *Phys. Rev. Lett.* 84, 951–954 (2000).
71. Jaramillo, T. F. et al. Identification of active edge sites for electrochemical H_2 evolution from MoS_2 nanocatalysts. *Science* 317, 100–102 (2007).
72. Somorjai, G. A. & Yang, M. The surface science of catalytic selectivity. *Top. Catal.* 24, 61–72 (2003).
73. Weissermel, K. & Arpe, H.-J. *Industrial Organic Chemistry*. 4th edn (Wiley-VCH, 2003).
74. Lefort, T. E. Process for the production of ethylene oxide. French Patent 729952 (1931).
75. Brainard, R. L. & Madix, R. J. Surface-mediated isomerization and oxidation of allyl alcohol on Cu(110). *J. Am. Chem. Soc.* 111, 3826–3835 (1989).
76. Linic, S. & Barteau, M. A. Formation of a stable surface oxametallacycles that produces ethylene oxide. *J. Am. Chem. Soc.* 124, 310–317 (2002).
77. Linic, S. & Barteau, M. A. Construction of a reaction coordinate and a microkinetic model for ethylene epoxidation on silver from DFT calculations and surface science experiments. *J. Catal.* 214, 200–212 (2003).
78. Linic, S., Jankowiak, J. & Barteau, M. A. Selectivity driven design of bimetallic ethylene epoxidation catalysts from first principles. *J. Catal.* 224, 489–493 (2004).
79. Lemons, R. A. Fuel-cells for transportation. *J. Power Sources* 29, 251–264 (1990).

80. Alayoglu, S., Nilekar, A. U., Mavrikakis, M. & Eichhorn, B. Ru-Pt core-shell nanops for preferential oxidation of carbon monoxide in hydrogen. *Nature Mater.* 7, 333–338 (2008).
81. Liu, P., Logadottir, A. & Nørskov, J. K. Modeling the electro-oxidation of CO and H_2/CO on Pt, Ru, PtRu and Pt_3Sn. *Electrochim. Acta* 48, 3731–3742 (2003).
82. Studt, F. et al. Identification of non-precious metal alloy catalysts for selective hydrogenation of acetylene. *Science* 320, 1320–1322 (2008).
83. Shustorovich, E. & Bell, A. T. The thermochemistry of C-2 hydrocarbons on transition-metal surfaces - the bond-order-conservation approach. *Surf. Sci.* 205, 492–512 (1988).
84. Kovnir, K. et al. A new approach to well-defined, stable and site-isolated catalysts. *Sci. Technol. Adv. Mater.* 8, 420–427 (2007).
85. Volpe, M. A., Rodriguez, P. & Gigola, C. E. Preparation of Pd-Pb/α-Al_2O_3 catalysts for selective hydrogenation using $PbBu_4$: the role of metal-support boundary atoms and the formation of a stable surface complex. *Catal. Lett.* 61, 27–32 (1999).
86. Choudhary, T. V., Sivadinarayana, C., Datye, A. K., Kumar, D. & Goodman, D. W. Acetylene hydrogenation on Au-based catalysts. *Catal. Lett.* 86, 1–8 (2003).
87. Blankenship, S. A., Voight, R. W., Perkins, J. A. & Fried, J. E. Process for selective hydrogenation of acetylene in an ethylene purification process. US Patent 6,509,292 (2003).
88. Kohan, A. F., Ceder, G., Morgan, D. & van de Walle, C. G. First-principles study of native point defects in ZnO. *Phys. Rev. B* 61, 15019–15027 (2000).
89. Solans-Monfort, X., Branchadell, V., Sodupe, M., Sierka, M. & Sauer, J. Electron hole formation in acidic zeolite catalysts. *J. Chem. Phys.* 121, 6034–6041 (2004).
90. Pacchioni, G. Modeling doped and defective oxides in catalysis with density functional theory methods: Room for improvements. *J. Chem. Phys.* 128, 182505 (2008).
91. Chretien, S. & Metiu, H. O_2 evolution on a clean partially reduced rutile TiO_2(110) surface and on the same surface precovered with Au_1 and Au_2: The importance of spin conservation. *J. Chem. Phys.* 129, 074705 (2008).

92. Eder, F. & Lercher, J. A. Alkane sorption in molecular sieves: The contribution of ordering, intermolecular interactions, and sorption on Brønsted acid sites. *Zeolites* 18, 75–81 (1997).
93. Dion, M., Rydberg, H., Schroder, E., Langreth, D. C. & Lundqvist, B. I. Van der Waals density functional for general geometries. *Phys. Rev. Lett.* 92, 246401 (2004).
94. Chakarova-Kack, S. D., Schroder, E., Lundqvist, B. I. & Langreth, D. C. Application of van der Waals density functional to an extended system: Adsorption of benzene and naphthalene on graphite. *Phys. Rev. Lett.* 96, 146107 (2006).
95. Johannesson, G. H. *et al.* Combined electronic structure theory and evolutionary search for materials design. *Phys. Rev. Lett.* 88, 255506 (2002).
96. Curtarolo, S., Morgan, D., Persson, K., Rodgers, J. & Ceder, G. Predicting crystal structures with data mining of quantum calculations. *Phys. Rev. Lett.* 91, 135503 (2003).
97. Oganov, A. R. & Glass, C. W. Crystal structure prediction using ab initio evolutionary techniques: Principles and applications. *J. Chem. Phys.* 124, 244704 (2006).
98. Newns, D. M. Self-consistent model of hydrogen chemisorption. *Phys. Rev.* 178, 1123–1135 (1969).
99. Kitchin, J. R., Nørskov, J. K., Barteau, M. A. & Chen, J. G. Modification of the surface electronic and chemical properties of Pt(111) by subsurface 3d transition metals. *J. Chem. Phys.* 120, 10240–10245 (2004).
100. Jacobsen, C. J. H., Dahl, S., Boisen, A., Clausen, B. S. & Nørskov, J. K. Optimal catalyst curves: Connecting DFT calculations with industrial reactor design and catalyst selection. *J. Catal.* 205, 382–387 (2002).

Chapter 4

Study and Application of Numerical Simulation of Deep Profile Control with Weak Gel

Zhou Yazhou and Yin Daiyin

Institute of Petroleum Engineering, Northeast Petroleum University, China

ABSTRACT

With the development of numerical reservoir simulation technology, numerical simulation has been widely applied to studies on the mechanics of all chemical flooding and injection approaches. In this study, a mathematical model for three-dimensional two-phase six-component deep profile control with weak gel is established based on the mechanics of deep profile control with weak gel. The established model takes diffusion, crosslinking reaction, adsorption, degradation and other physical or chemical phenomena into consideration. A numerical simulation software for deep profile control with gel was created with FOR99 programming language. Using this software, we performed optimization on the slug combination and slug size for the intra-flooding deep profile control with gel in ASP flooding for Xingbei development zone of Daqing oil field. The analysis on the mechanics of deep profile control with gel demonstrated that the addition of the profile control system could strengthen the profile adjustment effect. After the crosslinking system is injected into oil layer, less displacing fluid could enter the high-permeability layers, stimulating more displacing fluid to enter moderate or low-permeability layers, thereby the oil displacement effect is improved.

INTRODUCTION

Deep profile control with weak gel is a new technology integrated with the characteristics of polymer flooding. Weak gel combined the characteristics of conventional water shutoff profile control and polymer flooding, it can modify and improve the heterogeneity of the formations in deep reservoirs in order to redirect the fluid and expand the swept volume [1-3]. Meanwhile, weak gel can be used as displacing phase to improve the unfavorable mobility ratio of water flooding, thereby increasing the oil displacement efficiency of the injected water. The profile control ability of weak gel is embodied in its macromolecules, which could improve both horizontal and vertical heterogeneities of reservoirs, adjust the permeability difference between the water-adsorbing profiles and the reservoir, redirect the subsequent fluids, and expand the swept volume [4-5]. The flooding mechanism of weak gel: by increasing the water viscosity and improving the mobility ratio of water flooding, weak gel can improve the displacing efficiency and eventually increase the oil recovery rate. In this paper, we will introduce a numerical simulation method for deep profile control with weak gel. A mathematical model of deep profile control with weak gel is established, descriptions of related physical and chemical mechanisms are given, and the difference method that is stable and fast is adopted to solve related partial differential equations. With the established model, we conducted injection approach optimization for intra-flooding deep profile control with gel in ASP flooding. This study provides guidance for the implementation of mining methods.

MATHEMATICAL MODEL FOR DEEP PROFILE CONTROL WITH WEAK GEL

Assumptions

- Rocks and Fluids are compressible;
- Seepage in the whole process is isothermal;
- The movement of fluids follows the Darcy's law;
- The diffusion in the process follows the Fick's law;
- Chemical reactions only occur between polymer and crosslinker;
- Impact of plugging agent on water-phase density can be ignored;
- Fluid is composed by oil phase, water phase, and six pseudo-components including oil, water, polymer, crosslinker, gel, and salt; oil phase only contains oil component, and the other components are contained in water phase;
- Balance among the phases is reached instantly;
- The adsorption of polymer, crosslinker, and gel on rock surfaces follows the Langmuir isothermal adsorption theory, and the process is irreversible;
- The residual resistance factor has influence on the permeability rates of both oil phase and water phase.

Establishment of the Mathematical Model for Deep Profile Control with Gel

Equation (1-6) are the mathematical models for each component established according to the law of conservation of mass.

Oil:

$$\nabla \left[\frac{\rho_o K K_{ro}}{R_k \mu_o B_o} (\nabla p_o - \gamma_o \nabla D) \right] + q_o = \frac{\partial}{\partial t} (\phi \rho_o S_o) \tag{1}$$

Water:

$$\nabla\left[\frac{\rho_w KK_{rw}}{R_k \mu_w B_w}(\nabla p_w - \gamma_w \nabla D)\right] + q_w = \frac{\partial}{\partial t}(\phi \rho_w S_w)$$

(2)

Polymer:

$$\nabla(C_p \frac{\rho_w KK_{rw}}{R_k \mu_w B_w}(\nabla p_w - \gamma_w \nabla D)) + \nabla(D_p \phi_p \rho_w S_w \nabla C_p) + \phi \rho_w S_w R_{pre} + q_p = \frac{\partial}{\partial t}(\phi_p \rho_w S_w C_p + \rho_r(1-\phi)C_{pad})$$

(3)

Crosslinker:

$$\nabla(C_c \frac{\rho_w KK_{rw}}{R_k \mu_w B_w}(\nabla p_w - \gamma_w \nabla D)) + \nabla(D_c \phi \rho_w S_w \nabla C_c) + \phi \rho_w S_w R_{cre} + q_c = \frac{\partial}{\partial t}(\phi \rho_w S_w C_c + \rho_r(1-\phi)C_{cad})$$

(4)

Gel:

$$\nabla(C_g \frac{\rho_w KK_{rw}}{R_k \mu_w B_w}(\nabla p_w - \gamma_w \nabla D)) + \nabla(D_g \phi_g \rho_w S_w \nabla C_g) + \phi \rho_w S_w \{R_{gre} + R_{gke}\} = \frac{\partial}{\partial t}(\phi_g \rho_w S_w C_g + \rho_r(1-\phi)C_{gad})$$

(5)

Salt:

$$\nabla(C_s \frac{\rho_w KK_{rw}}{R_k \mu_w B_w}(\nabla p_w - \gamma_w \nabla D)) + \nabla(D_s \phi \rho_w S_w \nabla C_s) + q_s = \frac{\partial}{\partial t}(\phi \rho_w S_w C_s)$$

(6)

Where $\lambda_a = \frac{K_a}{\mu_a}$, $a = 0$ or w, ϕ_g are the accessible pore volume coefficients of polymer and gel, C_p, C_c, C_g and C_s are the concentrations of polymer, crosslinker, gel, and salt, D_p, D_c, D_g and D_s are the diffusion coefficients of polymer, crosslinker, gel, and sal, Rpre, Rcre and R_{gre} are the concentrations of polymer, crosslinker, and gel consumed or produced by crosslinking reaction in a time unit, C_{pad}, C_{cad} and C_{gad} are the adsorption amounts of polymer, crosslinker, and gel on rock surfaces.

The auxiliary equations are:

Saturation equation:

$$S_o + S_w = 1$$

(7)

Capillary pressure equation

$$P_{cow} = P_o - P_w = P(S_w, \sigma_{wo})$$

(8)

PROCESSING OF PARAMETERS

Water Viscosity

Viscosity without Considering Shear Rate

The influence of threshold crosslinking concentration is considered for the processing of the viscosity of weak gel system [6-7]. When the concentration of the crosslinking system is lower than threshold crosslinking concentration, the empirical formula proposed by Flory-Huggins should be used; when the concentration of the injection system is higher than the threshold, the modified formula of Flory-Huggins (Equation (9)) should be used.

$$\mu_{gel}^0 = \begin{cases} \mu_w\left(1+b_1 C_p + b_2 C_p^2\right) & C_{gel} \le C'_{gel} \\ \mu_w\left[1+b_1 C_p + b_2 C_p^2 + b_3\left(C_{gel} - C'_{gel}\right)^3\right] & C_{gel} > C'_{gel} \end{cases} \quad (9)$$

Where μ_{gelo} is the viscosity of the salt-less and shear-less crosslinking system, C'_{gel} is the threshold crosslinking concentration, b_1, b_2 and b_3 are coefficients.

Viscosity of Salt-contained and Shearing Injection System

Quite a few factors would influence the viscosities of polymer and crosslinking system, for example the relative molecular weight, temperature, salinity, and mechanical shearing of polymer and inter crosslinking system polymer. The Meter equation (Equation (10)) describes the influence of shearing on the viscosity of polymer solution.

$$\mu_p = \mu_\infty + \frac{\mu^0 - \mu_\infty}{1+\left(\dfrac{\gamma}{\gamma_{0.5}}\right)^{n-1}} \quad (10)$$

Where μ^0 the viscosity of the polymer solution without shear rate is μ_∞, is the viscosity when the shear rate is infinite γ, is the shear rate, is the shear rate $\gamma_{0.5}$ when $\mu = 0.5\mu^0$, n is the power law exponent of non-Newtonian fluid, and $1.0 \leq n \leq 1.8$.

When the influences of salinity and shear rate are considered:

$$\gamma = \frac{268|u|}{[KK_{rw}/(\phi S_w)]^{1/2}}$$

The viscosities of polymer and gel are expressed by Equation (12) and (13), respectively.

$$\mu_p = \mu_w + \left(\mu_p^{0_e - a_1 C_{se}} - \mu_w\right) e^{-a_2(1+a_3 C_{se})\gamma} \quad (12)$$

$$\mu_{gel} = \mu_{po} + \left(\mu_{gel}^{0_e - b_1 C_{se}} - \mu_{po}\right) e^{-b_2(1+b_3 C_{se})\gamma} \quad (13)$$

Where μ_p is the viscosity of the polymer solution when the influences of salinity and mechanical shearing are considered, μ_{gel} is the viscosity of the crosslinking system solution when the influences of salinity and mechanical shearing are considered, C_{se} is the effective salinity, and a_1, a_2, a_3, b_1, b_2 and b_3 are coefficients.

By linear summation of the viscosity of each component with Equation (14), the viscosity of the compound solution could be obtained.

$$\mu_T = f(\mu_w, \mu_p, \mu_{gel})$$

(14)

Relative Permeability Rates of Oil and Water Phases

The relative permeability rates of oil and water phases can be calculated by Equation (15) and (16).

$$K_{ro}(S_w) = K_{ro}(S_{wc}) \cdot (\frac{1-S_w-S_{or}}{1-S_{wc}-S_{or}})^{n_o} \tag{15}$$

$$K_{rw}(S_w) = K_{rw}(S_{or}) \cdot (\frac{1-S_w-S_{wc}}{1-S_{wc}-S_{or}})^{n_w} \tag{16}$$

Calculate the logarithms of the two sides of the two equations above, substitute the measured actual relative permeability rates of oil phase and water phase, and the straight line slopes obtained after linear regression are the relative permeability rate factors n_o and n_w of oil phase and water phase, respectively.

Adsorption

The adsorption or retention of polymer and gel on the rock surfaces are the main factors reducing the stratums permeability [8]. The factors influencing the adsorption are mainly the type, concentration, and molecular size of the polymer and gel, as well as parameters of the stratum environment like temperature, salinity, rock type, clay content, permeability, and porosity. Assuming the adsorption of chemical agents is irreversible and follows the Langmuir adsorption law, the adsorption could be expressed by Equation (17-19).

$$C_{pad} = C_{padmax} \frac{a_1 C_p}{1+b_1 C_p} \tag{17}$$

$$C_{cad} = C_{cadmax} \frac{a_2 C_p}{1+b_2 C_p} \tag{18}$$

$$C_{gad} = C_{gadmax} \frac{a_3 C_p}{1+b_3 C_p} \tag{19}$$

Where q_{pad}, q_{cad} and q_{gad} are the maximum adsorption capacities of polymer, crosslinker, and gel on rock surfaces, a_1, a_2, a_3, b_1, b_2 and b_3

are the adsorption equilibrium constants whose values are measured in labs.

Accessible Pore Volume

Since polymer and gel are both macromolecules, the produced gel can only pass through the pores with larger throats, and the pores with smaller throats are inaccessible. The porosities of the pores accessible by polymer and gel are and ϕ_p, respectively; ϕ_g and the inaccessible pore volume is defined as equation (20) [9].

$$IPV = \frac{\phi - \phi_a}{\phi}$$

(20)

$$\phi_a = (1 - IPV)\phi$$

Then
(21)

Where a=p or a=g.

Residual Resistance Coefficient

The polymer in the gel solution would adsorb on the surface of the pores in oil layers and retain inside them, which would cause the decrease of the permeability rate of water phase. The residual resistance coefficient R_k is a measurement of the decrease in the permeability rate caused by the gel produced after the polymer solution flows through porous media and develops chemical reactions, its expression is as shown in Equation (22) [10].

$$R_k = 1 + \frac{(R_{k\max} - 1)b_{rk} \times C_p}{1 + b_{rk} \times C_p}$$

(22)

Where $R_{k\max}$ is the maximum residual resistance coefficient, b_{rk} is an undetermined constant.

Crosslinking Reaction

The crosslinking reaction between polymer and crosslinker could be expressed as [11]:

Polymer (P) + Crosslinker (C) Gel (G)

Under certain reaction conditions, the following laws of chemical reaction dynamics could describe the crosslinking reaction rate [12-13]:

$$\frac{1}{C_p^0} \cdot \frac{dC_p}{dt} = -REC \cdot (C_p)^m \cdot (C_c)^n$$

(23)

$$\frac{1}{C_c^0} \cdot \frac{dC_c}{dt} = -REC \cdot (C_p)^m \cdot (C_c)^n$$

(24)

The reaction rates of polymer, cross linker, and gel are as expressed by Equation (25), (26), and (27), respectively.

$$R_{pre} = -R_e \left(C_p^0\right)^{m+1} \left(C_c^0\right)^n \cdot \left[1 + (m+n-1)R_e \left(C_p^0\right)^m \left(C_c^0\right)^n t\right]^{\frac{m+n}{m+n-1}}$$

(25)

$$R_{cre} = -R_e \left(C_p^0\right)^{n+1} \left(C_c^0\right)^n + 1\left[1 + (m+n-1)R_e \left(C_p^0\right)^m \left(C_c^0\right)^n t\right]^{\frac{m+n}{m+n-1}}$$

(26)

$$R_{pre} = -R_e \left(C_p^0 + C_c^0\right)\left(C_p^0\right)^m \left(C_c^0\right)^n \left[1 + (m+n-1)R_e \left(C_p^0\right)^m \left(C_c^0\right)^n t\right]^{\frac{m+n}{m+n-1}}$$

(27)

Where R_e is a constant for chemical reaction rate, its value is determined in labs or laws of chemical reaction dynamics, m and n are chemical reaction series, and $m \geq 1$, $n \geq 1$.

Degradation

The degradations of gel are mainly mechanical shearing, and chemical or thermal degradation [14]. Shear degradation mainly occurs because of the variation in velocity when the solution passes through pumps, pipes, perforations or enters the ground. Chemical degradation mainly refers to the inter-molecule chain breaking caused by the oxidation of the solution in stratums. Thermal degradation is decided by the thermal stability of polymer and gel in stratums.

For gels, the degradation rate R_{gde} could be expressed by Equation (28) [15].

$$R_{gde} = \frac{dc_g}{dt} = -K_{gde} \cdot C_g$$

(28)

Where K_{gde} is a degradation constant whose value is determined in labs

Definite Conditions of the Model

Boundary Conditions

The outer boundary of the model is closed boundary, i.e., $\frac{\partial \varphi}{\partial n}\big|_{t=0} = 0$ or constant pressure boundary. The inner boundary is line source (water injection well) and line congruence (production well). The boundary conditions of production well and water injection well are constant pressure or constant production.

Initial Conditions

Initial conditions of the model could be the initial conditions when the reservoir is commissioned, or the oil-phase pressure, water-phase saturation, and the mass concentration distributions of each component

when the reservoir is exploited into the end of a certain stage, $C_p\big|_{t=0}$

$= 0, C_g|_{t=0} = 0, C_c|_{t=0} = 0, C_s|_{t=0} = C_s$ (effective salinity of the original water in the stratum).

SOLVING OF THE MATHEMATICAL MODEL

The numerical discretization adopts the "finite difference" method to solve the mathematical model. The mass balance equations of each component are expressed by a series of discrete finite difference equations. The implicit solutions of these discrete non-linear simultaneous equations could obtain stable and longtime steps, and adopting adaptive implicit method could accelerate the simulation time and lower the demands on memory.

IMPES method is to solve the variables in the equation set step-by-step. Usually, the pressure should be calculated first with implicit method, and then the saturation is calculated with explicit method. The detailed process is described as below.

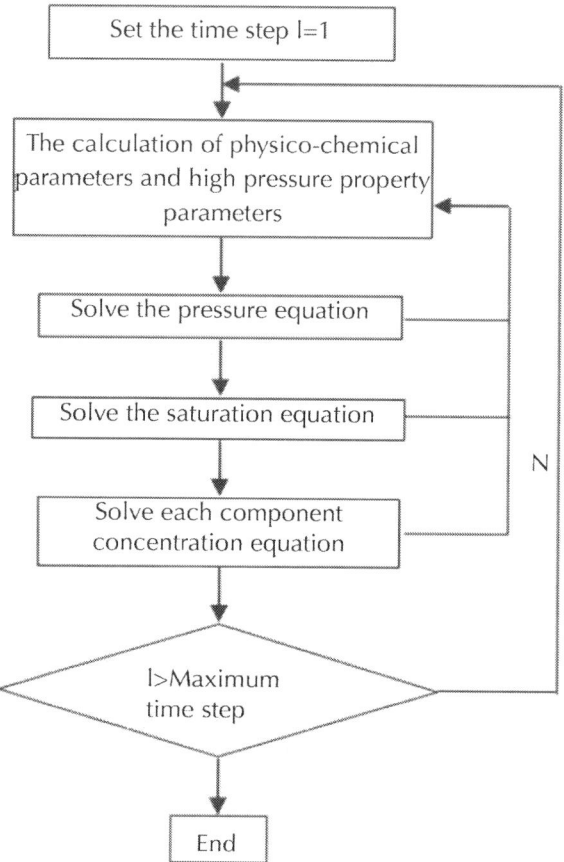

Figure 1: The thought of IMPES method solving.

APPLICATION OF NUMERICAL SIMULATION FOR DEEP PROFILE CONTROL WITH GEL

During the initial ASP slugging stage of the industrialized ASP flooding block of Xingbei development zone of Daqing oil field, due to the incompatibility of the injection technology and process, the performance of the injection system is poor. After process reformation and continuous follow-up adjustments, the response situation of the block is improved to a certain extent. However, unbalanced response situation is still outstanding in the block, and some production wells have bad response situations. This type of wells concentrates near the water injection well of the basic well pattern, where macro-pores exist at the bottom oil layer and low-efficiency or invalid circulations are serious. The high-permeability layers of these wells have relative fluid adsorption of up to 69.2%, while the fluid adsorption thickness ratio is only 23.9%; the injection strength of these wells reaches 15.6m^3/d.m, which is 9.4m^3/d.m greater than the average injection strength of the whole block. In addition, the oil layer growth of the injection wells in the block is predominantly full zone growth, with low potential of separated injection, thus deep profile control is needed. However, currently there is not any successfully profile control precedent in alkali system. Therefore, profile control technology in polymer flooding needs to be studied in order to guide the development of profile control approach for ASP flooding, improve the utilization of chemical agents, and enhance the overall production of the block. In-door experiments have helped optimize the formula of deep profile control with gel, and using numerical simulation technology, we performed optimization on the injection approaches.

Geological Model

In order to simulate the development effect of intra-flooding deep profile control with gel in ASP flooding, we established an ideal model using the software we created. The geological model is vertically composed by three layers with permeability of 300×10-3μm2, 500×10^{-3}μm^2 and 800×10^{-3}μm^2 from top to bottom, forming a positive rhythm reservoir. From top to bottom, the effective thicknesses of the layers are 1.5m, 2m, and 2.5m, the porosities of the layers are 0.26, 0.28, and 0.3, and the oil saturation of the layers are all 0.74. A Cartesian coordinate grid system is established, with 21 grids on both X and Y direction, step length of grid is 10.6m. Five wells are distributed in the oil layers in a 5-spot area well pattern; four of the wells are injection wells and one is production well, and the distance between injection wells and the production well is 150m. The plane grid division is as shown in Figure 2.

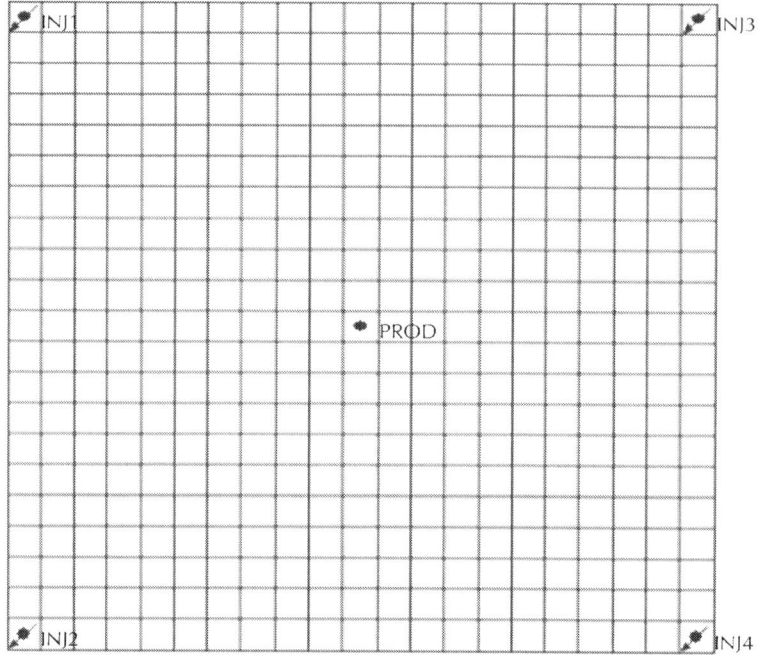

Figure 2: The grid subdivision schemes of geological model.

Optimization of Injection Parameters for Deep Profile Control with Gel

With the established ideal model, we performed optimization on the size of injected gel and the slug combination, and then compared the development effects of the optimized approach, water flooding, and ASP flooding. The injection rate of both water flooding and ASP flooding is 0.2PV/a, and the injection rate of gel is 0.18PV/a. The ASP formula is: polymer with concentration of 2000mg/L, NaOH with concentration of 1.2%, and surfactant with concentration of 0.3%.

Optimization of Slug Combination

In the site construction of ASP flooding, the injected alkali has a large concentration, usually 1.2%wt. This way, part of the alkali in the injected ASP flood fluid would adsorb on rock surface. When gel is injected afterward, these alkalis would weaken the performance of the gel, and the gel front would contact the ASP fluid, which would also weaken its performance. By taking these two types of impacts into consideration, we added polymer pre-slug at the gel front and then performed numerical simulation. The polymer pre-slug could dilute the residual alkali in the stratum, prevent the gel from contacting the ASP fluid, thereby ensuring good performance of the gel in the profile control process. Meanwhile, we also designed trail-slug and analyzed its influence on the development effect of deep profile control with gel.

Influence of the Size of Polymer Pre-Slug on the Development Effect

Polymer pre-slugs of 0.005PV, 0.01PV, and 0.015PV were designed, the approaches are detailed below.

Approach 1: water flooding until Water cut reaches 95% + 0.2PV of ASP + 0.035PV of gel + 0.2PV of ASP + post water flooding until Water cut reaches 98%;

Approach 2: water flooding until Water cut reaches 95% + 0.2PV of ASP + 0.005PV of polymer (with concentration of 2000mg/L) + 0.035PV of gel + 0.2PV of ASP + post water flooding until Water cut reaches 98%;

Approach 3: water flooding until Water cut reaches 95% + 0.2PV of ASP + 0.01PV of polymer (with concentration of 2000mg/L) + 0.035PV of gel + 0.2PV of ASP + post water flooding until Water cut reaches 98%;

Approach 4: water flooding until Water cut reaches 95% + 0.2PV of ASP + 0.015PV of polymer (with concentration of 2000mg/L) + 0.035PV of gel + 0.2PV of ASP + post water flooding until Water cut reaches 98%.

The numerical simulation result of the four approaches is as shown in Table 1.

Table 1: Development effects of the approaches

Approach	Recovery rate (%)	Recovery rate increment comparing to recovery rate without polymer pre-slug (%)
Approach 1	73.77	/
Approach 2	74.08	0.31
Approach 3	74.33	0.56
Approach 4	74.39	0.62

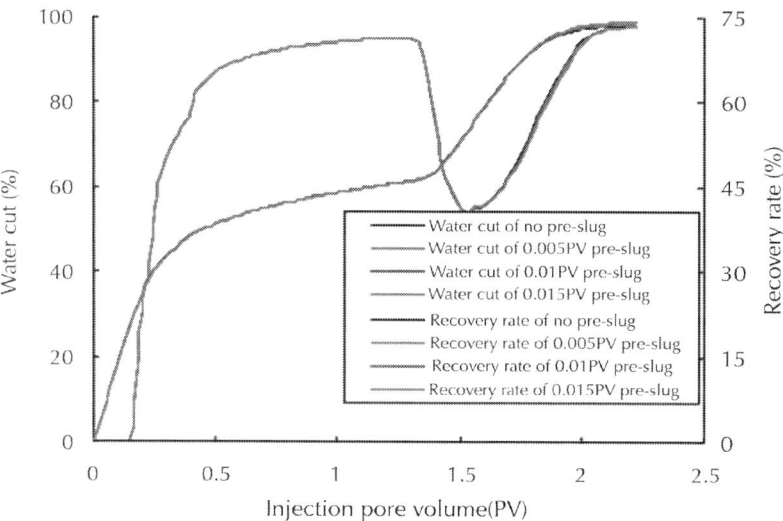

Figure 3: Comparison of recovery rate and water cut of each approach.

It can be seen that with the increase of the size of the polymer pre-slug, the recovery rate increases as well; however, after the size of the polymer pre-slug reaches 0.01PV, the increase in the size has little influence on the recovery rate. When there is no polymer pre-slug, the recovery rate is 73.77% when the Water cut reaches 98%. When polymer pre-slug of 0.005PV is added, the recovery rate increased by 0.31%. When the size of the polymer pre-slug is increased to 0.01PV, the recovery rate further increased by 0.25%. When the size is increased to 0.015PV, the recovery rate further increased by 0.06% only. Considering from economic and development effect views, the polymer pre-slug of 0.01PV is chosen.

Influence of the Concentration of Polymer Pre-slug on the Development Effect

Based on the optimization of polymer pre-slug, pre-slug of 0.01PV is chosen. Three approaches with concentrations of polymer slug as 1500mg/L, 2000mg/L, and 2500mg/L are designed to analyze the influence of the concentration of polymer pre-slug on the development effect of deep profile control with gel. The approaches are detailed below.

Approach 1: water flooding until Water cut reaches 95% + 0.2PV of ASP + 0.01PV of polymer (with concentration of 1500mg/L) +0.035PV of gel +0.2PV of ASP + post water flooding until Water cut reaches 98%;

Approach 2: water flooding until Water cut reaches 95% + 0.2PV of ASP + 0.01PV of polymer (with concentration of 2000mg/L) +0.035PV of gel +0.2PV of ASP + post water flooding until Water cut reaches 98%;

Approach 3: water flooding until Water cut reaches 95% + 0.2PV of ASP + 0.01PV of polymer (with concentration of 2500mg/L) +0.035PV of gel +0.2PV of ASP + post water flooding until Water cut reaches 98%.

The numerical simulation result of the three approaches is as shown in Table 2.

Table 2: Development effects of the approaches

Approach	Recovery rate (%)	Recovery rate increment comparing to recovery rate without polymer pre-slug (%)
Approach 1	74.06	0.29
Approach 2	74.33	0.56
Approach 3	74.37	0.60

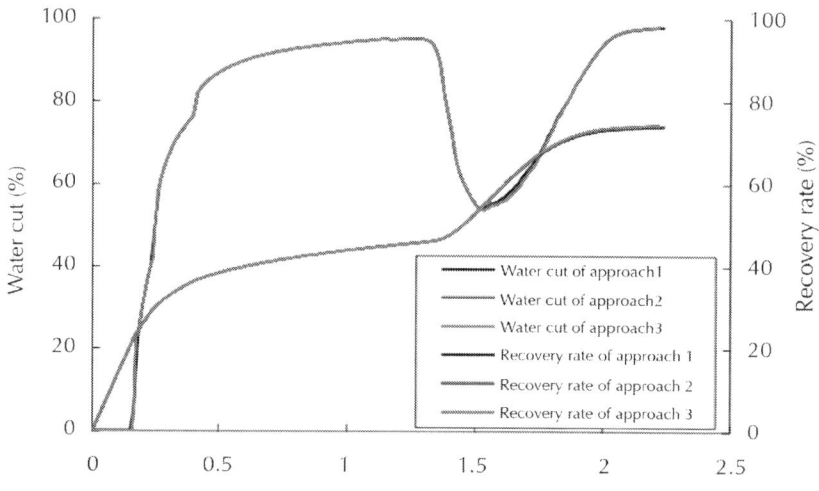

Figure 4: Comparison of recovery rate and water cut of each approach.

It can be seen that that with the increase in the concentration of the polymer pre-slug, the recovery rate increases as well. When the concentration is 1500mg/L, the recovery rate is 74.06%, which is 0.29% higher than the recovery rate without polymer pre-slug; when the concentration increases to 2000mg/L, the recovery rate increased by 0.35% comparing with that when the concentration is 1500mg/; when the concentration is 2500mg/L, the recovery rate further increased by 0.04% only. Overall considering from economic and effective perspectives, the concentration of polymer pre-slug is determined as 2000mg/L.

Combining the analysis in the two sections above, the concentration of the polymer pre-slug for deep profile control with gel determined as 2000mg/L, and the size of the pre-slug is decided as 0.01PV.

Influence of Trail-slug on the Development Effect

Based on the previous analysis, trail-slug is added on the basis of pre-slug with the concentration of 2000mg/L and the size of 0.01PV, two approaches are designed for the analysis of the influence of trail-slug on the effect of deep profile control with gel.

Approach 1: water flooding until Water cut reaches 95% + 0.2PV of ASP + 0.01PV of polymer (with concentration of 200mg/L) +0.035PV of gel +0.2PV of ASP + post water flooding until Water cut reaches 98%;

Approach 2: water flooding until Water cut reaches 95% + 0.2PV of ASP + 0.01PV of polymer (with concentration of 2000mg/L) +0.035PV of gel +0.2PV of ASP + 0.01PV of polymer (with concentration of 2000mg/L) + post water flooding until Water cut reaches 98%.

The numerical simulation result of the three approaches is as shown in Figure 5.

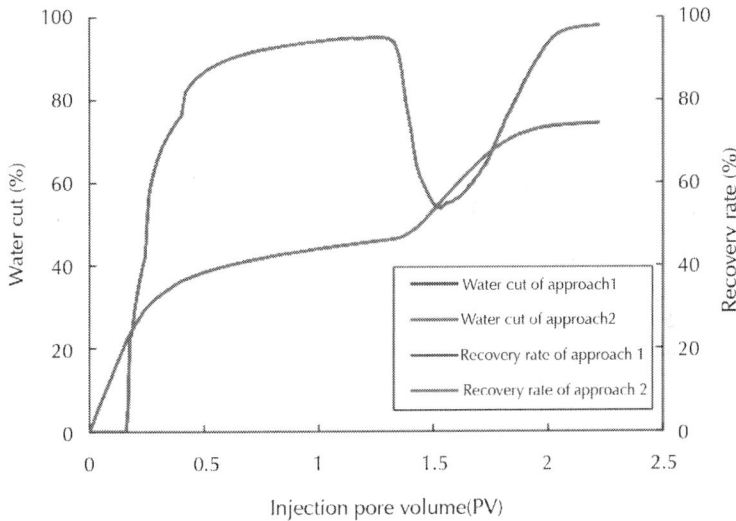

Figure 5: Comparison of recovery rate and water cut of each approach.

It can be seen that the addition of 0.01PV of trail-plug has little influence on the effect of gel deep profile control. The recovery rate without trail-plug when the Water cut reaches 98% is 74.33%, while the recovery rate with 0.01PV of trail-plug is 74.46%. The increment of 0.13% has little influence on the effect of deep profile control with gel.

Based on the above analysis, pre-slug has significant influence on development effect, whereas trail-plug has little. The reason is that the polymer pre-slug dilutes the residual alkali in the stratum and prevents the gel from contacting the ASP fluid, thereby ensuring good performance of the gel in the profile control process. Therefore, the optimal slug combination for gel flooding is to add 0.01PV of polymer solution with concentration of 2000mg/L before gel flooding.

Slug Size Optimization for Deep Profile Control with Gel

Three approaches with different sizes of slugs are designed as below.

Approach 1: water flooding until Water cut reaches 95% + 0.2PV of ASP + 0.01PV of polymer (with concentration of 200mg/L) +0.02PV of gel +0.2PV of ASP + post water flooding until Water cut reaches 98%;

Approach 2: water flooding until Water cut reaches 95% + 0.2PV of ASP + 0.01PV of polymer (with concentration of 200mg/L) +0.035PV of gel +0.2PV of ASP + post water flooding until Water cut reaches 98%;

Approach 3: water flooding until Water cut reaches 95% + 0.2PV of ASP + 0.01PV of polymer (with concentration of 200mg/L) +0.05PV of gel +0.2PV of ASP + post water flooding until Water cut reaches 98%.

The simulation result of the three approaches is as shown in Table 3.

Table 3: Development effects of the approaches

Approach	Recovery rate (%)	Recovery rate increment comparing to ASP flooding (%)
Approach 1	72.95	3.20
Approach 2	74.33	4.58
Approach 3	74.74	4.99

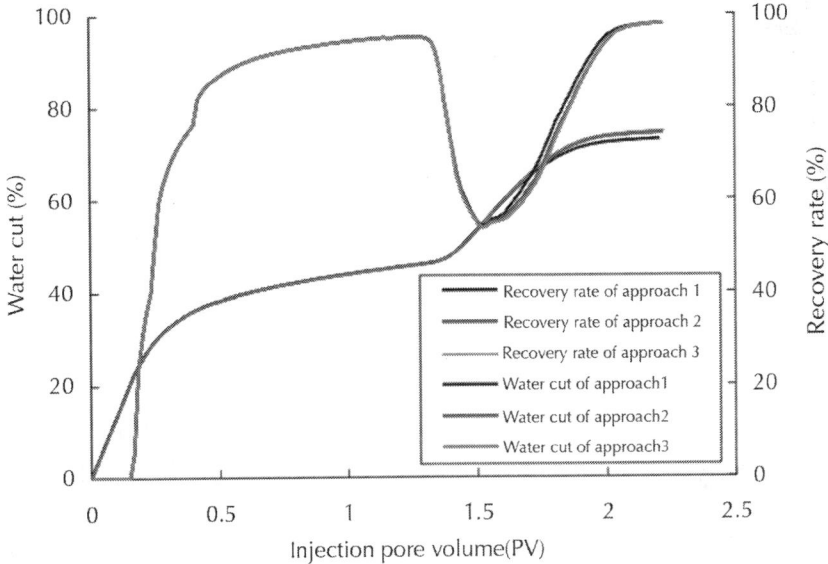

Figure 6: Comparison of recovery rate and water cut of each approach.

It can be seen that when of the slug size of injected gel increases, the recovery rate increases as well; however, after the size of the slug reaches 0.035, the increase has little influence on the recovery rate. When the slug size is 0.02PV, the recovery rate increased by 3.20% comparing to ASP flooding; when the slug size is 0.035PV, the recovery rate further increased by 1.38%; when the slug size is 0.05PV, the recovery rate only further increased by 0.41%. Therefore, gel slug of 0.035PV is the optimal choice.

Based on the analysis on the slug combination and slug size optimization for deep profile control with gel, polymer pre-slug of 2000mg/L and 0.01PV+0.035PV of gel is the optimal choice for intra-flooding deep profile control with in ASP flooding.

Evaluation of Deep Profile Control with Gel Effects

In order to evaluate the profile control effects of gel in ASP flooding, we compared the deep profile control with gel, water flooding, and ASP flooding with the following specifics.

Water flooding: Water flooding until Water cut reaches 98%;

ASP flooding: Water flooding until Water cut reaches 95% + 0.4PV of ASP + post water flooding until Water cut reaches 98%;

ASP flooding with deep profile control of gel: Water flooding until Water cut reaches 95% + 0.02PV of ASP + 0.01PV of polymer (with concentration of 2000mg/L) + 0.035PV of gel + 0.2PV of ASP + post water flooding until Water cut reaches 98%.

The numerical simulation result of the three flooding approaches is as shown in Table 4.

Table 4: Development effects of the three flooding approaches

Approach	Lowest Water cut (%)	Decrement (%)	Recovery rate (%)	Recovery rate increment comparing to water flooding (%)
Water flooding	/	/	51.33	/
ASP flooding	56.25	38.75	69.75	18.42
ASP and gel	53.84	41.16	74..33	23.00

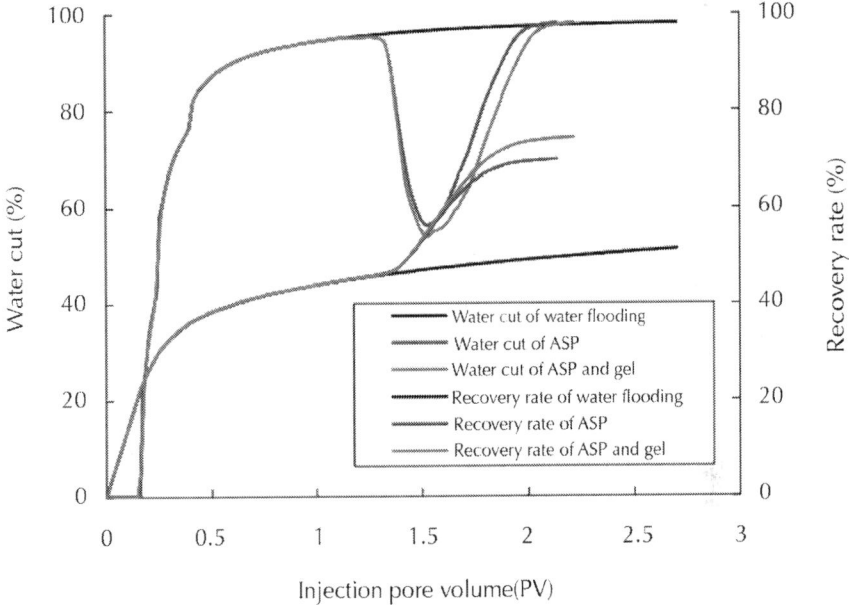

Figure 7: Comparison of development effects of the three oil displacement approaches.

According to the numerical simulation result, the addition of 0.035PV of deep profile control with gel in ASP flooding could increase the recovery rate of water flooding by 23%, and increase that of ASP flooding by 4.58%. The lowest Water cut is 53.84%, which reduced by 41.46% comparing with the Water cut before chemical flooding, and 2.41% comparing with the Water cut before ASP flooding.

Analysis of the Mechanics of Deep Profile Control with Gel

In order to study the mechanics of deep profile control with gel, we analyzed the fluid production, Water cut, and post-displacement oil saturation of individual well at high, moderate, and low permeability, respectively. For layers of different permeability in individual well, Figure 8-10 present the calculated fluid production, Figure 11-13 present the calculated Water cut, and Figure 13-15 present the calculated post-displacement oil distribution.

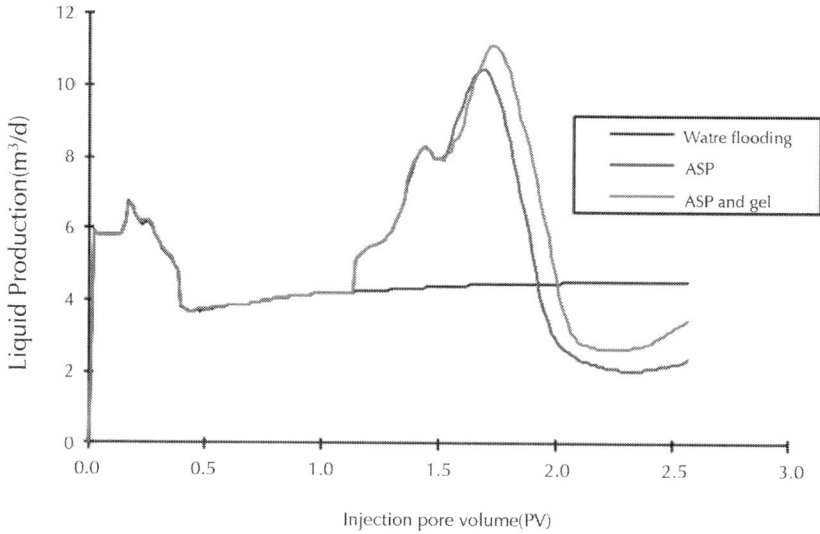

Figure 8: Fluid productions in low-permeability layers.

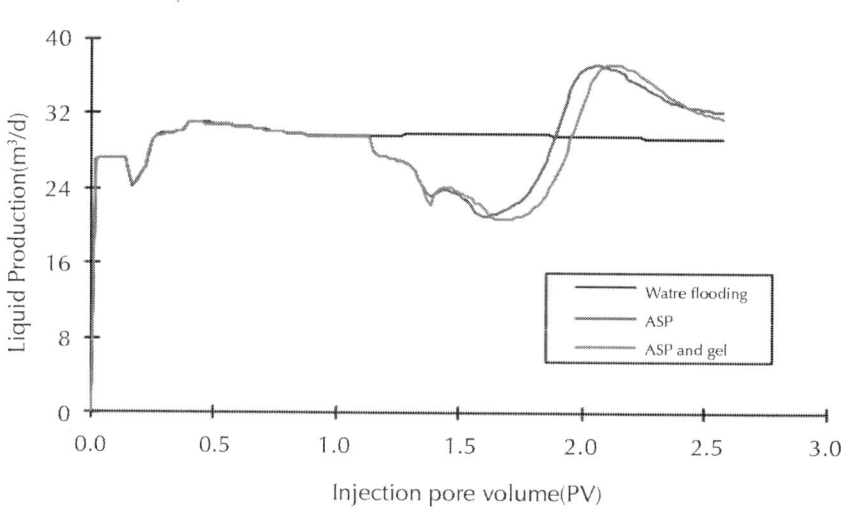

Figure 9: Fluid productions in moderate-permeability layers.

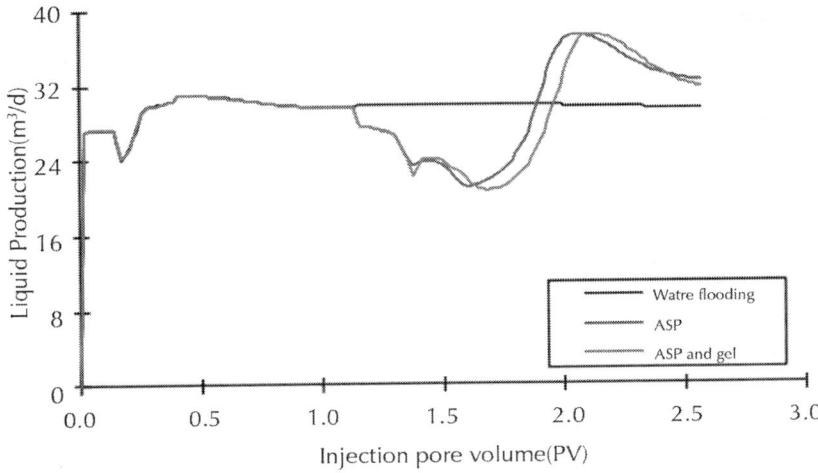

Figure 10: Fluid productions in high-permeability layers.

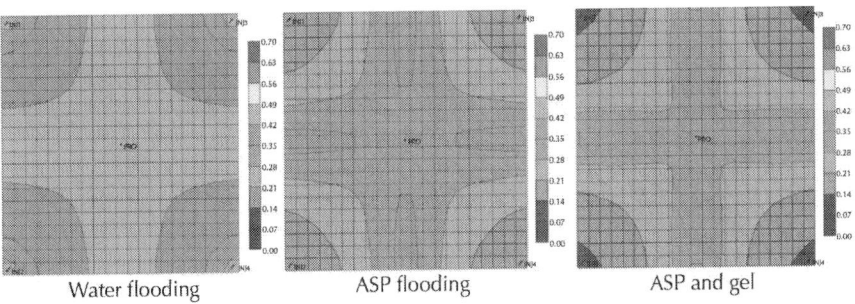

Figure 11: Post-displacement oil distributions of three oil displacement approaches in low-permeability layers.

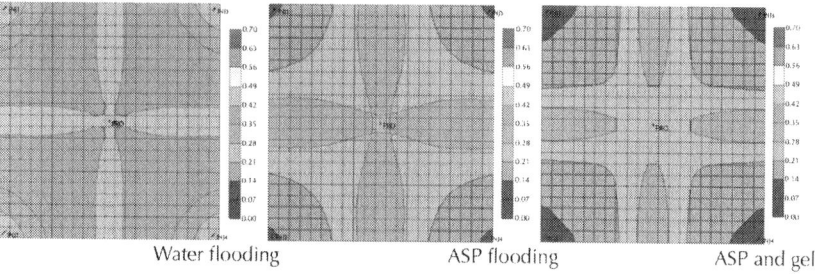

Water flooding   ASP flooding   ASP and gel

Figure 12: Post-displacement oil distributions of three oil displacement approaches in moderate-permeability layers.

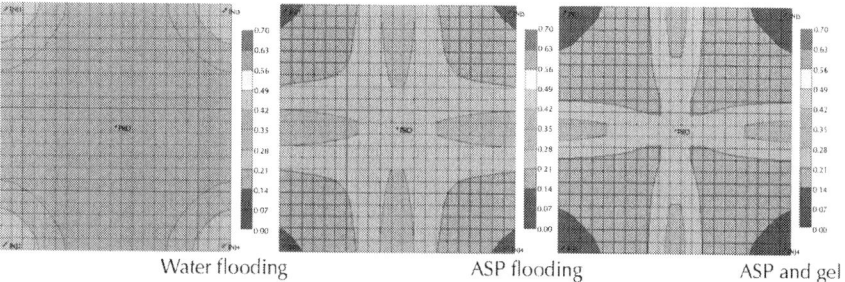

Water flooding   ASP flooding   ASP and gel

Figure 13: Post-displacement oil distributions of three oil displacement approaches in high-permeability layers.

It can be seen from Figure 8-10 that serious channeling of the displacing fluid along high-permeability layers occurred in water flooding due to the unfavorable mobility ratio, the moderate- and low-permeability layers are not displaced well. For ASP flooding, the increase of water-phase viscosity adjusted the profile to a certain extent; comparing with water flooding, the fluid adsorption of high-permeability layers decreased because of the increase in flow resistance, forcing the displacing fluid to enter the moderate- and low-permeability layers, which improves the oil displacement effect in these layers. After adding deep profile control with gel in ASP flooding, the profile adjustment effect is improved, and the high-permeability layers are better slugged; when the crosslinking system is injected to the oil layer, the displacing fluid that enters the high-permeability layers is reduced, stimulating more displacing fluid to enter moderate- and low-permeability layers, thereby improving the oil displacement effect. Seen from Figure 11-13, the oil saturation of low-permeability layers barely changed after water flooding, which means the oil is basically not displaced, whereas the oil in high-permeability layers are well displaced. As for ASP flooding, the increase of the viscosity of the injected fluid has certain profile adjustment ability, which improved the producing degree of the oil in low-permeability layers. After adding 0.035PV of deep profile control with gel in ASP flooding, the producing degree of low-permeability layers is further improved. According to the analysis of the simulation result, the deep profile control with gel in ASP flooding could improve the profiles and enhance the oil recovery rate.

CONCLUSIONS

- In this study, a mathematical model for three-dimensional two-phase six-component deep profile control with weak gel was established based on the mechanics of deep profile control with weak gel. This model takes diffusion, crosslinking reaction, adsorption, degradation and other physical or chemical phenomena into consideration. A numerical simulation software for deep profile control with gel was created with FOR99 programming language.
- Based on the geological features of the ASP flooding areas of Daqing oil field, an ideal model was established using the created software. With this model, we performed optimization on the slug combination and slug size for the deep profile control with gel in the ASP flooding areas of Daqing oil field and obtained the optimal injection approach. Our study showed that polymer pre-slug has greater influence on the development effect, while trail-slugs have little. The reason is that the polymer pre-slug would dilute the residual alkalis in the stratum and prevent the gel from contacting the ASP fluid, so that the good performance of gel during the profile controlling process is ensured.
- The created software was used for the analysis of the mechanisms of deep profile control with gel. The analysis demonstrated that in ASP flooding, the employment of deep profile control with gel could improve the profile modification effect, and strengthen the plugging of high-permeability layers. After the crosslinking system is injected into oil layer, less displacing fluid could enter

the high-permeability layers, forcing more sweeping fluid to enter moderate or low-permeability layers, thereby the oil displacement effect is improved. After the addition of deep profile control with gel, the producing degree of low-permeability layers is further improved.

ACKNOWLEDGEMENTS

This work is supported by the National Natural Science Foundation of China under Grant No.51074034.

REFERENCES

1. D. Y. Yin, H. Pu and Y. X. Wu, "Numerical simulation of imbibition oil recovery for low permeability fractured reservoir", Chinese Journal of Hydrodynamics, vol. 4, (2004), pp. 19.
2. D. Y. Yin and H. Pu, "Numerical Simulation Study on Surfactant Flooding for Low Permeability Oilfield in The Condition of Threshold Pressure", Journal of Hydrodynamics, vol. 4, (2008), pp. 20.
3. D. Y. Yin, Y. G. Guo and W. M. Huang, "The determination of reasonable cycle time for bailing oil production in low permeability field", The 21st session of the national seminar on hydrodynamics and the 8th national conference on hydrodynamics and on both sides of the ship and ocean engineering hydrodynamics seminar, (2008) August 562-567; China.
4. H. Q. Jiang, W. Liu, M. Yuan and M. R. Sun, "Numerical Simulation Research on Times Profile Control", Oil Drilling & Production Technology, vol. 5, (2003), pp. 25.
5. W. Liu, "The Numerical Simulation Research of Multi round Profile Control in the Reservoir of High Water", Chinese Journal of Engineering Mathematics, vol. 4, no. 22, (2005), pp. 580-584.
6. G. Chen, G. Zhao and Y. L. Ma, "Mathematical Model of Polymer Linked Profile Control Enhanced Oil Recovery", Journal of Tsinghua University (Science and Technology), vol. 12, no. 44, (2004), pp. 1607-1609.

7. G. Zhang, Q. H. Feng, D. K. Tong, Y. Liu, S. D. Liu and Y. B. Chen, "Mathematical Model of Mobile Gel for Deep Profile Control and Oil Displacement and Fast Solution Methods", Petroleum Geology and Recovery Efficiency, vol. 04, no. 15, (2008), pp. 56-58.
8. X. C. Wu, W. Y. Zhu, Q. K. Ma, Y. Zhang, H. Chen and Y. P. Liu, "Study on Nonlinearity Seepage Characteristic and Mathematical Model of Movable Gel", Oil Drilling & Production Technology, vol. 28, no. 05, (2006), pp. 43-45.
9. Q. H. Feng, G. Zhang, Y. Tao and W. D. Wu, "Study of Fast Simulation Method for Flowing Gel Driving", Journal of China University of Petroleum (Edition of Natural Science), , vol. 6, no. 30, (2006), pp. 64-70.
10. B. Xu and L. S. Cheng, "Numerical Simulation of Deep profile control with Weak Gel", Chinese Journal of Computational Physics, vol. 2, no. 22, (2005), pp. 164-170.
11. Y. Kyung, J. -w. Kim, S. -B. Jung and K. -H. Eom, "Optimal Control Method for a Hydroelec-tric Power Development in Multi-Level Dams", International Journal of Control and Automation, vol. 3, no. 13, (2009).
12. M. Li and Z. J. Kang, "Research and Application of Numerical Model of Profile Control In-depth", Petroleum Geology & Oilfield development in Daqing, vol. 3, no. 23, (2004), pp. 77-80.
13. K. P. Song, "Fluid Poromechanics in Polymer Flooding", Beijing: Petroleum Industry Press, (1999).
14. T. L. Chen, X. J. Zhou, F. P. Tang and S. Y. Wang, "Recovery Enhance Technology with Deep profile control with Weak Gel", Beijing: Petroleum Industry Press, (2006).
15. R. D. Kokate and L. M. Waghmare, "Review of Tuning Methods of DMC and Performance Evaluation with PID Algorithms on a FOPDT Model", International Journal of Control and Automation, vol. 4, no. 95, (2011).

# Chapter 5

Environmental Impact Assessment of Oil and Gas Sector: a Case Study of Magurchara Gas Field

J. B. Alam[1], A. A. M. Ahmed[2], G. M. Munna[1], and A. A. M. Ahmed[3]

[1]Department of Civil and Environmental Engineering, Shahjalal University of Science and Technology, Sylhet-3114, Bangladesh

[2]Department of Civil Engineering, Leading University, Sylhet-3100, Bangladesh

[3]Department of Business Administration, Leading University, Sylhet-3100, Bangladesh

ABSTRACT

This study focuses on the environmental impact assessment of Magurcherra gas field through environmental, socio-economical and meteorological study. The major activities involved are seismic activities, drilling activities, exploration and production. In the case of Magurcherra gas field, improper planning and drilling activities have created explosion and caused huge environmental damage. In order to evaluate the environmental damage due to exploration and drilling activities, checklist of possible environmental impact on the nature, society and socioeconomic activities of the proposed area has been identified by focus group discussion, survey and sample testing. During the observation pollution related with noise found as temporary and minor but the air is observed to be polluted by dust, SPM, SO_x, NO_x, and CO_2 etc. The effects of explosion on the natural forests, land, wildlife, tea garden, infrastructure, local people and tea garden workers, livestock feeding, communication and transport etc are measured. During observation, water quality deterioration by spills and leaks of oil and grease, paints, solvents and chemicals was identified. It was found from the cumulative analysis of the study that the project is environmentally feasible.

INTRODUCTION

Bangladesh is poor in natural resources by global standards. Natural gas is the most important natural resource that has been discovered and being utilized. Other important discoveries include coal, peat, hard rock and modest reserves of limestone, gravel, glass sand and various types of clay. Environmental issues and problems in Bangladesh are now recognized as the key concerns for the sustainable development of the country. Magurcherra gas field, 8 km from Srimangal on the road to kamalganj. The gas field caught fire in 1997 and was ablaze for three months, laying waste to betel-nut plantations and tea estates in the vicinity (McAdam, 2008). Generally, in the petroleum sector major activities involved are seismic activities, drilling activities, exploration and production. During those activities the Surrounding areas are affected in various ways. During drilling operations, drilling fluid or mud is continuously circulated through drill pipe and back to the surface equi-pments to balance underground hydrostatic pressure, to cool the bit and flush out rock cuttings. The location of a drill site depends on the characteristics of the underlying geological formation; however, environmental impacts can be minimized by selecting appropriate site for drilling.

One of the earliest contaminant of the drilling site is the mud used while drilling. Huge quantities of drilling mud are required continuously. The returned mud is recycled, but finally to be disposed of the effluent mud is usually stored in large pits. Around four to five thousand cubic meters of mud is stored at a time. Eventually water evaporates, leaving behind the clay in the pit and contaminates top soil. Drilling-mud requires large quantities of water from the nearby sources. It is necessary to check whether the water is contaminated by pathogens. If the water is contaminated, it must be treated. Large quantitiesof fluid are let of on the ground or into a nearby stream. The amount of effluent disposed will be very damaging for the neighboring areas. It is particularly damaging in the case of off-shore exploration as there is no alternative but to dispose it off into the sea (BUET, 2003). Environmental Impact Assessment (EIA) for all petroleum projects will allow the planners and implem-enters to understand the environmental impacts and con-sequences of their projects (NEAB, 1998; DoE, 2001).

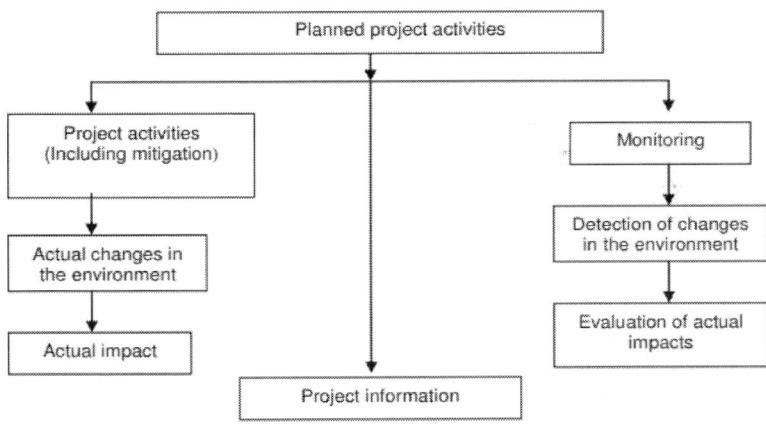

Figure 1: Flow diagram for EIA of the project.

The EIA process is thought to consist of six main components: basics; impact identification; description of the affected environment; prediction and assessment of impacts; selection of proposed action and documentation in accordance with extant guidelines (Center, 1997). EIA has been followed to define a project and development alternatives, the environment on which it may be affecting, the potential impacts, mitigation alternatives, impact monitoring and public interests. In order to find out possible beneficial and adverse impacts, environmental impact analysis, construction related impacts, air pollution, thermal pollution of waste resources, noise pollution, soil cont-amination, sediment and dust pollution, waste disc-harges, occupational health and safety, changes in aesthetic environment etc were carried out in this study. The major objectives of the present study were (i) to assess the socio-economic impacts of these projects, (ii) to assess the impacts on land use pattern and (iii) to identify and quantity the environmental effects of these projects.

METHODOLOGY

Checklist method was used in this study. A questionnaire was developed and the following assessment was made based on expert's opinion. 100 experts from different fields related to gas and oil sector were asked and the evaluation was prepared. Pollutants emission rates from the power plant are calculated by using the fuel consumption rate and emission factor (Kato and Akimoto, 1992) for the unit consumption. The Gaussian Plume Model (Peterson, 1978) was used to estimate the pollutant concentration. Environmental Impact Assessment (EIA) was assessed by the graded matrix system developed by Leopold et al. (1971) in which 'magnitude' and 'importance' of the impact in each cell of a matrix can be denoted by assigning numerical values. The flow diagram of the EIA project shown in Figure 1:

Baseline Environmental Condition of Magurchara Gas Field

Magurchara Gas Field Explosion

The gas field caught fire in 1997 and caused huge damage of life and property over the whole area. The Magurchara gas field explosion damaged about 60 hectors of natural forest and 300 hectors of land were burnt. A large part of wild life (Deer, birds, foxes, monkeys etc) was destroyed or displaced to other places. Around 3000 people were affected because 31 hectors of tea garden was completely damaged during explosion (EPCT, 1997). So, the overall socio-economic and environmental scenario was highly affected by the gas field explosion. Preliminary assessment of environmental damages and deterioration of Magurchara Gas Field Explosion are given in Table 1 and its effects on plants, soil and atmosphere are shown in Figure 2 which shows the deterioration soil properties, effects on plants on the Magurchara gas field site.

Noise

Noise, which is seldom recognized as a source of pollution, constitutes a danger to people's health through physical, physiological and even psychological stresses. Drilling operations in particular are very noisy and pose a threat to the relevant workers at the site.

Production of Formation Water

Oils produced from wells are invariably accompanied with water as they are in close association in oil pool. Hence, whenever oil is produced, some quantity of water is also produced. On an average up to 20% water is associated with oil. This water needs to be separated and disposed off in a suitable manner. This water is usually charged with droplets of oil and some salts. Droplets of oil are collected as far as practicable by using demulsifies. Sometime this water is used for injecting into the water-horizon in the oil field. The quantity of water to be disposed of is colossal and requires careful thoughts as to how it can be done so that the environment is not damaged.

Table 1: Preliminary environmental assessment of Magurchara gas field explosion

Resources	Component	Preliminary assessment
	Natural forest	60 ha. completely damaged
		100 ha completely burnt
	Land (300 ha.)	Partly burned observed
Natural resources		Covered with ash and condensate observed
		Landslide / Land subsidence observed
	Wildlife	Deer, birds, foxes, monkeys etc. destroyed or displaced
	Water logging/ pollution	Observed
	Tea Garden	31 ha. completely damaged
		1 km Train line damaged
Development resources	Infrastructure	1 km Medalled road, 2 culverts damaged
		1 km Gas line
	Livestock	Not observed
	Fisheries	Not observed
	Local people and Tea garden workers	Around 3000 people affected
	Livestock feeding	Not observed
	Communication and transport	Disrupted
Human interest	Socioeconomic disruption	Observed
	Air pollution	Not measured
	Population / market displacement	Not observed

Environmental Impact Assessment of Oil and Gas Sector: a Case Study...

Figure 2: Effects of Magurchara gas field explosion on the environment.

Table 2: Possible environmental impacts of pipeline construction

Environmental component	Positive and negative impact	Mitigation measures
Socio-economic	Negative	Providing temporary housing, eating and sanitary facilities for the construction force to prevent overtaxing the local infrastructure
Land use	Negative	Compensation to owners have to be made for crop loss, land should generally be acquired by individual agreement with the owners
Soil fertility	Negative	Soil fertility is to be preserved by segregating the 30 cm. topsoil layer from common fill material during trenching.
Air quality	Impact is negligible (No)	No mitigation measures are necessary

Surface and groundwater quality	Negative	Potable water used by the construction force
		have to be tested to ensure that it meets the
		quality standards of Bangladesh for drinking
		water. Implementation of Waste Disposal Plan
		including proper sanitary facilities for the
		construction force and proper disposal of solid
		waste generated by the construction activities.
Fish and wildlife	Negative	Select alternate routes of pipeline to avoid
		forest. Construction force has to be prohibited
		from hunting to prevent further degradation of
		this limited resource. Natural fish production
		has to be protected by controlling water
		Pollution.
Historical and archaeology resources	Impact is considered minor (Negative	

RESULTS AND DISCUSSION

Environmental impact assessment of drilling and pipe line activities was predicted for the Magurchara Gas Field exploration in oil and gas sector. A questionnaire was developed and the following assessment was made based on expert's opinion. 100 experts from different fields related to gas and oil sector were asked and the evaluation was prepared. In Table 2, Negative and Positive were used to classify the magnitude of the environmental parameters with the relevant mitigation measures. The impact on the socio-economic, land use, surface and groundwater quality, fish and wild life is stated as negative where the air quality is negligible. A minor impact is stated on historical and archaeological resources.

Environmental evolution has been shown in Table 3 where the identified environmental parameters are analyses for finding out the feasibility of the project by Leopold matrix. Leopold et al. (1971) have developed graded matrix system in which 'magnitude' and 'importance' of the impact in each cell of a matrix can be denoted by assigning numerical values. This approach is used of gross screening technique for impact identification purposes. This study used checklist method and identified the impact through three different dimensions of environment physical, ecological and socio-economic. The effects are stronger on the eco-logical and socioeconomic environment rather than physical environment. The effect on plane land is evaluated on low and hilly terrain land medium. Most of the components of physical environment are evaluated as insignificantly affected. Ecological environment of the study area is seriously affected by the gas field explosion. In the study area a huge amount of forests are distracted and got high gradein evaluation. The effects on wildlife and migrated birds are evaluated as low and medium, respectively. Distribution of wetland is also considered by the gas field. In socio-economic environment agriculture sector, crops and plantation, and farming are affected and the effects are evaluated as medium. The workers of the gas field and the irrigation are affected highly by the gas field explosion. Other important components of socio-economic environment like industrial, residential, commerce and industry, household, land communication, social structure are also affected and effects are evaluated as low.

Table 3: Environmental impact evaluation of Magurchara gas field explosion

S/No.	Environmental component	Negative Impact Probability (p)	Severity (s)	Impact Value (IV) = p * s	No Impact	Positive Impact	Insignificant	Low	Medium	High
1.	**Physical environment**									
	Topography									
	Plane land	1	6	6				√		
	Hilly terrain	2	6	12					√	
	Drainage									
	Congestion	2	2	4			√			
	Flash flood									
	Hazard									
	Earthquake	1	3	3			√			
	Cyclone/storm									
	Water contamination									
	Surface water	1	2	2			√			
	Ground water	1	1	1			√			
	Bio-chemical	2	2	4			√			
	Soil									
	Erosion	2	2	4			√			
	Siltration				√					
	Pollution	2	1	2			√			
	Air pollution									
	SPM, dust	3	2	6				√		
	So$_x$, No$_x$, CO$_2$	2	5	10					√	
2.	**Ecological environment**									
	Terrestrial Flora									
	A forestation	1	6	6				√		
	Destruction of plantation	3	6	18						√
	Aquatic flora									
	Eutrophication				√					
	Nuisance plant	1	3	3			√			
	Terrestrial Fauna									
	Disturbance to wildlife	4	3	12				√		
	Disturbance to migrated birds	3	4	12					√	
	Aquatic fauna									
	Destruction of wetland	2	3	6				√		
3.	**Socio-economic environment**									
	Loss of land									
	Agriculture	4	4	16					√	
	Residential/Community	2	2	4			√			
	Industrial/Commercial	2	3	6				√		
	Impact On									
	Crops/Plantation	2	6	12					√	
	Residential/Community	1	3	3				√		
	Commerce/Industry	1	3	3				√		
	Indirectly affected									
	Household	1	3	3				√		
	Transportation									
	Land communication	2	4	8				√		
	Health and Safety									
	Workers	3	6	12						√
	Local people	2	4	8				√		
	Noise to neighbors	4	2	8				√		
	Disturbance to									
	Social structure	4	2	8				√		
	Major sources									
	Irrigation	2	5	10						√
	Farming	2	4	8					√	

CONCLUSIONS

It is clear from the study that the cumulative impact of the project is +220 which indicates that the project is environmentally feasible. During back filling of trench the fertility of soil has been reduced and the water quality has been deteriorated by spills and leaks of oil and grease, paints, solvents and chemicals. The fish production of this area has been observed to be affected by water poll-ution from oil and grease, chemicals and sanitary wastes. Pollution related with noise found as temporary and minor but the air is observed to be polluted by dust, SPM, SO_x, NO_x, and CO_2 etc. Water used by the cons-truction force have to be tested to ensure that it meets the quality standards of Bangladesh for drinking water, or it should be chlorinated so that the tested chlorine residual is 0.2 mg/l or greater, after 10 min of contact time. Implementation of Waste Disposal Plan including, proper sanitary facilities for the construction force and proper disposal of solid waste generated by the construction activities. Soil erosion has to be minimized with the measures for conserving soil during stream crossings and during trenching activities.

REFERENCES

1. BUET (2003). Workshop in Oil and Gas Sector", organized by Bangladesh University of Engineering and Technology, Dhaka.
2. Canter LW (1997). "Environmental impact assessment", 2nd edition, Mcgraw-hill Inc., p. 640
3. DoE (2001). "EIA Guidelines", Department of Environment, Ministry of Environment and Forest, Govt. of Bangladesh
4. EPCT (1997). "Draft Report - Environmental Impact Assessment in Oil and Gas Sector", survey work conducted by Engineering Planning Consultancy Team, Sylhet.
5. Kato N, Akimoto H (1992). Anthropogenic Emissions of SO2 and NOx in Asia: Emission Inventories. Atmospheric Environment, 26A: 2997- 3017.
6. Leopold LB, Clarke FE, Manshaw BB, Balsley JR (1971). A Procedure for Evaluating Environmental Impacts, U.S. Geological Survey Circular No. 645, Government Printing Office, Washington, D.C.

7. McAdam M (2008). Bangladesh Travel Guide, Lonely Planet, 6th Edition ISBN: 9781741045475, p. 153.
8. NEAB (1998). "EIA Manual", National Environmental Association of Bangladesh.

CITATION

J. B. Alam, A. A. M. Ahmed, G. M. Munna, and A. A. M. Ahmed, Environmental impact assessment of oil and gas sector: A case study of Magurchara gas field, ISSN 2141-2391

Ghana's Quest for Oil and Gas: Ecological Risks and Management Frameworks

P. A. Sakyi[1], J. K. Efavi[2], D. Atta-Peters[1] and R. Asare[3]

[1]Department of Earth Science, University of Ghana Legon, Accra, Ghana

[2]Department of Material Sciences, Faculty of Engineering Sciences, University of Ghana,

PMB, Legon, Accra, Ghana

[3]Science and Technology Policy Research Institute, Council for Scientific and Industrial

Research Cantonments, Accra, Ghana

ABSTRACT

Ghana discovered commercial oil and gas in 2007, and, subsequently, commenced production in the last quarter of 2010. In the light of the potential economic boost that will accompany petroleum production, its discovery was welcome news for Ghanaians. However, oil exploration and production involve several activities that can have detrimental impacts on the ecosystem. In this paper, the potential sources of pollution in the upstream sector of the oil and gas industry and their effects on the environment are discussed. Also discussed are existing national environmental management legislations in the extractive industry, and the implementation and enforcement challenges these regulations face. Strategies to curtail the effects of oil and gas development on the ecosystem are also put forward. These include the need for government to formulate petroleum industry-specific environmental protection guidelines and appropriate regulatory frameworks. Such regulations in managing the environment should employ an integrated approach involving (i) prescription of environmental codes and setting of standards by government to be met by operators, and (ii) the need for oil companies to develop environmental management system (EMS) to ensure that they operate within the environmental standards for the industry. Administrative and institutional restructuring and reforms, as well as the provision of the necessary financial and human resources for the various environmental agencies, should be encouraged to ensure effective implementation, enforcement and monitoring.

INTRODUCTION

The global economy relies heavily on oil and gas to fulfill majority of its energy demands, and it is a key indicator of the economic wellbeing of both developed and developing nations. The International Energy Agency (IEA) predicts that global oil demand will reach 90 million barrels/day in 2010 and about 104 million barrels/day by 2020 (World Energy Outlook, 2009). The continued increase in the world's energy demands is due, in part, to robust economic growth in China and India and an uncertain political climate in the Middle East (Mane, 2005; BP, 2008) The west central coast of West Africa along the Gulf of Guinea is reported to be endowed with rich hydrocarbon reserves, a source of oil and gas. Production of oil and gas in this region, which is estimated to have over 547 offshore oil and gas structures, has the potential to meet the energy demands of the European Union and the United States of America (Ayoade, 2002; IMF, 2005).

Angola, Nigeria, Equatorial Guinea and Gabon are already producing crude oil in this region. Apart from the huge hydrocarbon reserves, countries along the Gulf of Guinea, including Ghana, are now of global interest due to the geopolitical and geographical locations (ease of shipping, less prone to pirate attacks, immune to ground political instability and wars), as well as the quality of crude oil. This is in sharp contrast with other areas, such as the Middle East, which is noted for wars, uprisings and other unfavourable activities.

Ghana's oil and gas exploration dates as far back as 1896, and the commercial discovery in the Jubilee Field in June 2007 (GNPC, 2009) was, without doubt, welcome news, as it can enhance revenue generation and job creation, and significantly improve the national economy. The Jubilee Field is located 60 km off the coast of Cape Three Points in the Western Region of Ghana (Fig. 1). It is situated in the Deepwater Tano and West Cape Three Points blocks of the Tano Basin (Fig. 2), which is one of the three offshore sedimentary basins in Ghana. The Field is jointly owned by a consortium of companies named the Jubilee Joint Venture and managed by Tullow Ghana Ltd. Water depth within the Jubilee Field Unit area ranges from 1,000 m to 1,700 m, and the discovered hydrocarbon is a 36.5o API Sweet Crude having a 1,000–1,200 gas-to-oil ratio and little sulfur content by international standards (Sunu-Attah, 2009). The reserve is estimated to hold about 800 million to 1.5 billion barrels of oil (Adjaye, 2009).

In spite of the socio-economic benefits that oil and gas production can bring to the nation, oil exploration and production involve several activities that can have detrimental impact, either directly or indirectly, on the environment and, therefore, require special attention in striking a good balance. Though commercial production started in the last quarter of 2010, pre-production activities required stringent environmental measures to ensure that the ecosystem is not damaged. There are some existing laws in Ghana with relevance to the mining and oil industries. However, there are no comprehensive environmental laws directed at the oil industry (Darko-Mensah, 2009). Unregulated activities by the oil industry can potentially destroy habitats and damage biodiversity, and this requires that Ghanaian policy makers take a closer look at national sustainability strategies that will ensure sustainable exploitation of the hydrocarbon without adverse effects on the ecosystem.

This paper, as a contribution to the upstream sector of oil development in Ghana, highlights most of the upstream activities in oil and gas operations, the sources of pollution and the inherent environmental issues, and their potential negative impact on ecosystems. It also discusses existing national management practices to protect the environment, and recommends legislations and management frameworks that could be employed to minimize ecological damage.

MATERIALS AND METHODS

The Ecology of Cape Three Points Coastal Areas

There are more than 90 lagoons along the entire coastline of Ghana, most of which are located in the central and western coastline, including the Cape Three Points area. These lagoons serve as habitats and nursery sites for a variety of fish, shrimps, mollusc, and crabs species (Armah *et al.*, 2004). The ecology of the Cape Three Points area is broadly divided into two, namely offshore and onshore (Armah *et al.*, 2004), which are described in detail below.

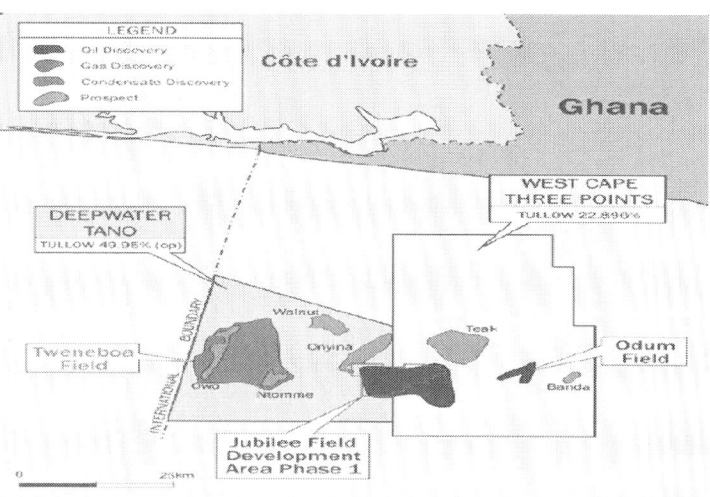

Figure 1: Offshore map of south-western Ghana, showing the location of the Jubilee Oil Field in the Deepwater Tano and West Cape Three Points blocks (Source: Asafo-Adjaye, 2011).

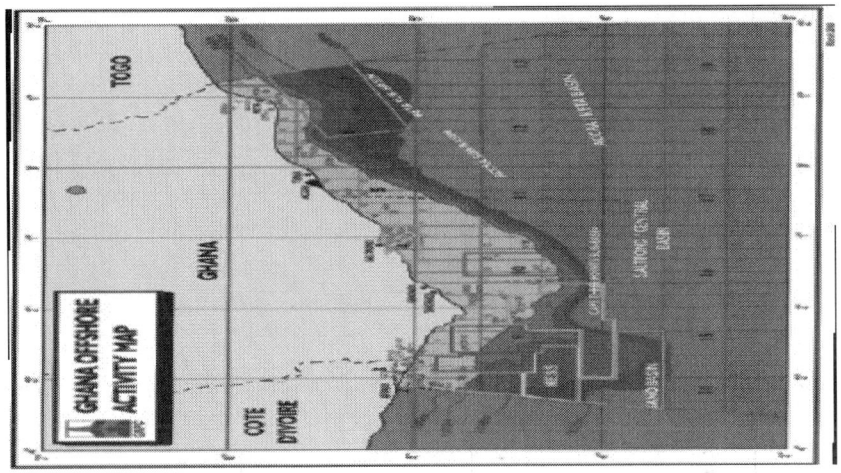

Figure 2: Ghana offshore activity map, showing the three offshore sedimentary basins in Ghana, namely the Tano Basin, Central (Saltpond) Basin and the Accra-Keta Basin (Source: Asafo-Adjaye, 2011).

The Offshore Ecology. An estimated total of 89 species of aquatic organisms are found in the 1,100–1,700 m water depth in the Jubilee Field. These include pelagic stocks and demersal species, which are the most abundant fish species offshore the Cape Three Points area. The pelagic stocks include sardinellas anchovy and chub mackerel (Boely & Freon, 1980). The distribution and abundance of the small pelagic species are affected by seasonal coastal upwelling in the area (MFRD, 2003). Bottom living demersal species in the area include red fishes, croakers, snappers, goatfish, groupers and threadfins. Other food species include shell fish, shrimps and prawns. Organisms of biomedical importance in this area include phytoplanktons, zooplanktons, pelagic fish, offshore benthic invertebrate fauna communities, marine mammals and sea turtles (MFRD, 2003).

Among the invertebrate fauna found in the coastal waters are polychaete worms, ribbon worms, amphipods, bivalves, gastropods, and decapod crustaceans (MFRD, 2003). In the Gulf of Guinea, the seasonal major and minor upwelling, occurring from June to September and December to January, respectively, each year, brings cold, nutrient-rich water to the surface, causing blooms of phytoplankton, which is dominated by diatoms such as *Leptocylindrons* sp., *Nitzchia* sp., *Chaetoceros* sp., *Rhizosolenia* sp. and *Skeletronema costatum*.

The Onshore Ecology. According to Armah *et al.* (2004), the sandy shores around the Cape Three Points region are known to serve as nesting places for sea turtles and habitats for species such as the ghost crab, isopods, amphipods, mysid, and mole crab. Others are the polychaetes, bivalves and gastropods. The sandy shores are also characterized by strand vegetation, including creepers and grasses, and changes further inland to dwarf palms, coconut palms and shrubs. The rocky shores, on the other hand, serve as important habitats for macro algae, barnacles and littorinid snails, all of which use the rocks as substrates (Armah *et al.*, 2004).

Along the coast of the Cape Three Points, which also hosts a forest reserve, rain forest fringes the Ehunli and Akpulu lagoons, and an estuary, where the Yile and Kpani rivers meet long the western coastline. A total of 141 species of 58 diverse plant families, made up of trees, shrubs, grasses and sedges, occur in this area (Armah *et al.*, 2004). In the forest are 263 species of birds, elephants, bongo, turtles and primates, including the endangered Diana monkeys. Because of the ecological importance of all the abovementioned species, it is important to regulate the activities of oil companies to ensure that they do not damage habitats and breeding grounds of the fauna and flora, which could cause them to be extinct.

Oil and Gas Development Activities

The oil and gas industry has three main sectors, namely the upstream, midstream and downstream. The upstream involves the exploration and production; the midstream covers the transportation of oil and gas, and the downstream deals with refining and processing of crude oil and gas products, as well as the distribution and marketing of the products (E&P Forum/UNEP, 1997).

The major stages of the upstream oil activities are briefly described below (E&P Forum/UNEP, 1997; Kloff & Wicks, 2004).

- Aerial and seismic surveys are carried out to identify favourable geological structures such as faults and anticlines in the subsurface.
- Exploration drilling and appraisal involve drilling of oil wells to confirm the presence or otherwise of hydrocarbon and the internal pressure of such a reserve, all aimed at evaluating the nature, size and extent of the hydrocarbon reservoir to enable confirmation of

its economic viability.
- Development and production wells are drilled into hydrocarbon reservoir to extract the produced fluids, comprising oil, gas and water.
- Decommissioning and rehabilitation involve the closure and removal of production installations and other structures at the end of the commercial life span of an oil reserve, followed by the restoration of the site to environmentally sound conditions.

Potential Environmental Impact Arising from Hydrocarbon Exploitation in the Jubilee Oil Field

As may be the case elsewhere, oil and gas exploration and production in the Jubilee Field involve the various stages that could be accompanied by intrinsic environmental challenges. The environmental impacts arising from oil and gas production activities can be broadly grouped into two, namely (i) ecosystems, and (ii) human, socio-economic and cultural (E&P Forum/UNEP, 1997). Of particular interest in this paper is the potential negative environmental impact on ecosystems. Accordingly, our discussion is focused on the ecological damage accompanying the upstream activities.

Noise. During oil and gas development, noise disturbances associated with aircraft, bulk vessels and drilling operational activities are likely to impact negatively on the ecosystem. This may arise from prospecting and survey activities already mentioned above. At certain levels, noise affects the functions of marine organisms. Fish and marine mammals, including whales and dolphins, are particularly affected mostly by sound elevation because of their dependence on sound for reproduction, feeding, and avoiding hazards such as predators and navigation (McCauley, 1994; Tyack & Miller, 2002; Popper, 2003). There have also been reported death, reduced growth, impaired hearing and stress, as some of the possible impact of noise from oil and gas operation (Fernandez *et al.*, 2005).

Atmospheric Emissions. Atmospheric emissions are increasingly becoming the subject of concern to both industry and national governments due to its negative effect on climate. Sources of emissions associated with oil development activities, can be grouped as follows (E&P Forum/UNEP, 1997):

- Flaring, venting and purging of gases;
- Combustion processes from diesel engines and gas turbines;
- Fugitive gases from loading operations and losses from process equipment;
- Airborne particulate from burning sources, such as well testing and soil disturbance during construction and vehicular traffic

Of these gas emissions, flaring is the most alarming, and has been a source of major conflict in Nigeria and elsewhere (Sala-i-Martin &Subramanian, 2003; ERA/CJP, 2005). The principal emissions accompanying flared gas contain toxic byproducts such as methane and benzene, and also generate carbon dioxide, carbon monoxide, volatile organic carbons, sulphur dioxide, nitrogen sulphide and nitrogen oxide. Some of these gases (e.g. carbon dioxide), contribute to global warming, whereas the sulphur gases and carbon dioxide contribute to the formation of acid rain, which is detrimental to soil fertility and vegetation upon interaction with water (Patin, 1999). Consequently, gas flaring has the potential to damage the Ankasa Forest Reserve and the surrounding vegetation and farmlands located offshore the Jubilee Field.

Current agreements between the Ghana Government and operators of the Jubilee Field emphasize on zero gas flaring. However, according to the GNPC and Tullow Ghana Ltd, there are no existing infrastructure to convert the natural gas into LPG to meet part of the country's energy demands (Ghana Oil Watch, 2011). On the other hand, re-injection of the gas back into the oil wells is not encouraged since that can damage the wells and reduce oil recovery, which, therefore, leaves the nation with the only option of flaring (Think Ghana, 2007). Currently ongoing is the Natural Gas Transportation and Processing Project (NGTPP), which is aimed at bringing natural gas from the Jubilee and the shallow water Tano fields and future discoveries for processing and further distribution to the Effasu Power Barge, the Takoradi Thermal Plant and for export (EDM/GNPC, 2009).

Aquatic Pollution. Discharges from oils and gas installations include produced water, process water, sewerage, sanitary and domestic wastes, and spills and leakages (E&P Forum/UNEP, 1997). These discharges arise from the drilling of exploration wells and, subsequently, the production of crude oil. Produced water is a combination of formation water from the reservoir and injection water, containing a complex mixture of inorganic and organic compounds, trace and heavy metals, drilling fluids and drill cuttings, and well treatment chemicals (E&P Forum, 1994; Sadiq et al., 2002). The composition of produced water makes it potentially toxic to marine waters.

Organic compounds in discharged waste water, when released into marine waters, rivers or lakes, react with and consume dissolved oxygen, thereby, depleting the water of oxygen and rendering it uninhabitable for aquatic organisms (Harremoës, 1998). Similarly, excess supply of nutrients to water bodies also stimulates excessive plant growth and causes reduction in water quality and a decrease in the population of fish and other aquatic organisms (Harremoës, 1998; WHO/EC, 2002). Anti fouling paints on ships also contain potent biocide such as tributyltin (TBT), which causes reproduction failure of female marine snails and a decline in population (Kloff & Wicks, 2004).

Oil tankers, underwater pipelines, offshore oil drilling rigs and coastal storage facilities can accidentally release crude oil into the ocean, and a significant portion of the ecosystem, both offshore and onshore Cape Three Points will potentially be at risk. The negative effects of oil spillage on marine organisms include damage to digestion tract of marine species through digestion, absorption of oil in contaminated food, contamination of eggs leading to poor hatchery, and trapping of turtles and birds leading to death. Over the years, the petroleum industry has witnessed oil spills that have caused considerable ecological damage.

Notable among these spills were the Amoco Cadiz, which spilled about 227,000 tonnes of oil in 1978 (Patin, 1999) and the Exxon Valdez, which spilled 40,000 tonnes of oil in 1989, resulting in the death of about 250,000 seabirds, nearly 3,000 sea otters, 300 harbour seals, 250 bald eagles and up to 22 killer whales (BBC, 1989). Similarly, in 1999, the Erika oil vessel spilled about 20,000 metric tonnes of oil that affected 400 km of coastline, and killed over 100,000 birds (BBC, 2000). The explosion, in 2010, of the Deepwater Horizon, owed by British Petroleum (BP) in the Gulf of Mexico killed 11 people, and resulted in the spillage of 4.9 million barrels of oil, polluting hundreds of miles of coastline and killing 491 birds, 227 turtles and 27 mammals within the first 40 days after the spill (Reuters, 2010; BBC, 2011).

Terrestrial Pollution. During oil and gas exploration and production, potential impacts on soils arise from physical disturbances due to construction, deforestation and contamination, resulting from spillage and leakage or solid waste disposal. These activities result in land degradation, transformation and fragmentation of natural habitats, and can disable the vital ecosystem processes that support growth (Barnard & Newby, 2009). In the Niger Delta region of Nigeria, three main sources of oil pollution have been identified, namely oil spills, gas flares and waste discharges (Pyagbara, 2007). Rivers, streams and ponds have been the receiving bodies for oil spills and waste discharges, with their accompanying negative environmental impacts.

Available data show that between 9 and 13 million barrels of oil have been spilt in the Niger Delta region in the past 50 years (NCF/WWF/IUCN, 2006). These spills, which occurred both on land and offshore, destroyed crops and damaged the quality and productivity of soil that the communities use for farming (UNEP, 2011). The spills have also caused the death of birds and mammals, damaged fisheries and contaminated water that the inhabitants use for drinking and other domestic purposes (Amnesty International, 2009).

Oil spills and other oil-related pollution have also seriously damaged the Niger Delta's mangroves, which are an important fish breeding area. The damage has resulted in a severely impaired coastal ecosystem, and compromised the livelihoods and health of the region's impoverished residents (NCF/WWF/IUCN, 2006; Amnesty International 2009), thus, negatively affecting economic activities. The reasons assigned to the frequent oil spills in the Niger Delta include corrosion of oil pipes, poor maintenance of infrastructure, spills or leaks during processing at refineries (World Bank, 1995), human error and the consequence of deliberate vandalism or theft of oil (Steiner, 2008). The damage to the ecosystem has caused the Ogoni people, who think their lives are intrinsically bound up with the survival of the environment, to stand up against the denigration of their environment (UNEP, 2011).

Ghana is likely to suffer from the abovementioned potential sources of pollution and their accompanying negative environmental impacts, if the environment is not well managed. Since the Jubilee Oil Filed is located offshore, the ecosystems of utmost concern are the ocean, beaches, and the atmosphere. The inhabitants of towns and communities dotted along the coast of the Gulf of Guinea in the Western Region of Ghana traditionally engage in fishing, as their means of livelihood. Consequently, protecting the sea from any potential environmental damage is very paramount.

Past Environmental Management Performance in Ghana – Case Study of the Mining Industry

Selected Mine-Related Pollutions In Ghana. Ghana has comprehensive legislation and regulations on environmental protection and supporting institutional infrastructure, like ministries, bureaus or agencies, all focused on the mining industry. These laws, notwithstanding, Ghana has experienced mining-related pollution of the aquatic and terrestrial ecosystems. Between 2001 and 2009, several mining communities in Ghana recorded more than nine cyanide spillages, most of which were caused by the collapse of tailings dams and holding ponds (e.g. EPA, 2004, 2005; GNA, 2009; Kosich, 2010). Notable among these occurred at the Newmont Ghana Gold Limited (NGGL) Ahafo Mine in 2009, which cost the company an amount of GH¢7 million in the form of fines and compensation to the Government of Ghana (Myjoyonline, 2010). The spillages damaged farmlands and polluted rivers, and, consequently, resulted in the death of thousands of fishes, crabs and shrimps, and posed health and environmental hazards to the people and wildlife in the respective communities.

In Ghana's oil and gas industry, besides the reported spillage of low toxicity oilbased mud by Kosmos Energy in the Jubilee Field in December 2009 and March 2010 (EPA, 2010), there are no documented incidents of spillages related to oil development in the coastal communities located offshore Saltpond and Cape Three Points, where oil is currently being produced. It is, therefore, important that the necessary measures are taken to ensure effective and efficient exploitation and production of oil and gas, thereby, minimizing or avoiding its attendant environmental consequences on the ocean and aquatic life.

Challenges in Implementation, Compliance, Enforcement and Monitoring.

Even though Ghana has well-formulated national policies and legal frameworks that regulate the operations of mining companies, most of the legislations on environment are not strictly enforced, and this has been attributed to several factors. Among these are weak institutional capacity to manage the environment, inadequate resources, and lack of political will, all of which have resulted in the lack of proper mechanisms for coordination, monitoring and enforcement. Furthermore, economic concerns, absence of effective sanctions to serve as deterrent to potential polluters, community dissatisfaction, and duplication and overlapping of institutional functions add up to the other foreseeable challenges (UNEP, 2002).

Finally, inadequate remuneration and lack of commitment on the part of staff members of the regulatory and enforcing agencies, often serve as good grounds for bribery and corruption. Consequently, mining companies find it cheaper to pollute than to prevent environmental degradation, and the consequence is the documented miningrelated pollution and land degradation in the mining communities.

Management Frameworks for Minimizing Oil related Ecological Risk

Legislations, Conventions and Regulatory Frameworks. Major environmental issues related to oil and gas development have been addressed through countless global and regional treaties, national laws and a number of administrative regulations and management frameworks, promulgated by individual countries and multinational organizations such as UN agencies, the World Bank, and International Finance Corporation (IFC) (Gao, 1998) to promote natural resource conservation and pollution control. Ghana is signatory to a number of United Nations and Regional Cooperation Conventions and multilateral agreements, which will help in managing environmental impacts. These international conventions are binding on national governments and serve as a baseline or guide in drafting national policies, legislations and *regulations*.

Notable among these treaties and conventions, that have been ratified by Ghana and of particular importance to the environment and oil and gas operations, include (e.g. Kloff & Wicks, 2004) 1. International Convention for the Prevention of Pollution of the Sea by Oil, 1962; 2.

International Convention on the Establishment of an International Fund for Compensation of Oil Pollution Damage, 1971; 3. The International Convention for the Cooperation in the Protection and Development of the Marine and Coastal Environment of the West and Central African Region, 1981 (Abidjan Convention); 4. International Convention on Civil Liability for Oil Pollution Damage, 1969; 5. Convention on Wetlands of International Importance, especially as Waterfowl Habitats, 1971; 6. Convention on the Conservation of Migratory Species of Wild Animals, 1979; 7. International Convention for the Conservation of Atlantic Tunas, 1966; 8. Montreal Protocol on Substances that Deplete the Ozone Layer, 1989; 9. Convention on Biological Diversity, 1992; 10. International Convention Relating to Intervention on the High Seas in Cases of Oil Pollution Casualties; 11. United Nations Convention on the Laws of the Sea, 1982; 12. The International Convention for the Prevention of Pollution from Ships (MARPOL Convention 73/78); 13. International Convention on Oil Pollution Preparedness, Response and Cooperation, 1990; 14. The Convention on the Control of Transboundary Movements of Hazardous Wastes and their Disposal (Basel Convention); 15. Convention on the Ban of the Import into Africa and the Control of Transboundary Movement of Hazardous Wastes within Africa (Bamako Convention).

Nationally, Ghana has no comprehensive environmental legislation targeting the oil and gas industry. Existing legislation for ecosystem protection include Wild Animals Preservation Act 1961 (Act 43), Oil in Navigable Waters Act, 1964 (Act 235), Wildlife Conservation Regulations 1971 (LI685), Wild Reserves Regulations 1971 (LI 740), Maritime Zones (Delimitation) Law (PNDCL 159 of 1986), Environmental Protection Agency Act 1994, The Wetland Management (Ramsar sites) Regulation, 1999, and Environmental Assessment Regulation 1999.

Other legal frameworks that target the oil and gas industry are Oil and Mining Regulations, 1957 (LI 221), Mineral (Offshore) Regulations 1963 (LI 257), Mineral (Oil and Gas) Regulations 1963 (LI 256), Ghana National Petroleum Corporation Act (Act 64 of 1983), Petroleum (Exploration and production) Law 1984 (PNDCL 84), and National Petroleum Authority Act (Act 691 of 2005). Some of these legislations were formulated for both the mining and oil and gas industries, and are, therefore, more generalized. Consequently, they have not been effective in the mining industry, and are, therefore, destined to face a number of challenges in the oil sector.

Recommended Measures to Address Environmental Concerns in the Oil and Gas Industry

Guided by the various international treaties and conventions, there is the urgent need for the Government of Ghana to formulate an all-inclusive oil and gas development policy with environmental issues at the centre stage. The policy framework should integrate environmental legislations and management systems, and also mandate stakeholders to develop an environmental value culture at every stage of their business processes to supplement government's efforts in a cost effective manner. Consequently, the policy should be tailored along two main approaches to regulating the environmental performance of an industry, namely the 'prescriptive' and 'performance based' approaches (Technical Meeting Document, 1998).

The Prescriptive Approach

The prescriptive or "command and control" approach is based on legislations indicating specific requirements made by government, to be met by operators. The regulations clearly spell out structural, technical, and procedural requirements to address environmental, health and safety hazards. This makes it relatively easy for government to determine, via an inspection procedure, whether an operator is meeting the requirements. Thus, it is convenient for the Government of Ghana to adopt this approach by setting mandatory environmental codes and standards to regulate and monitor the activities of companies in the oil and gas industry. These standards must include general guidelines for the preparation of an environmental impact assessment and detailed guidelines for the preparation of an environmental action/management plan to be submitted by firms before the commencement of operations. It is very important that environmental impact assessment be undertaken prior to the commencement of oil exploration and development, and, when discovered that it can potentially impact the environment negatively, the companies involved would be required to indicate what mitigation measures would be employed to contain the situation. The standards must also include acceptable limits of concentrations of compounds and chemicals in effluent discharges generated through the operations of the various companies. Equally importantly should be the application of the "polluter pays" principle to ensure that producers of wastes that cause environmental damage are made to pay compensation and the cost of remediation.

In line with this, a solid foundation has been laid to effectively manage the negative impact on the environment. First was a national forum held to discuss potential problems and solutions, including the environmental management of oil development. Also formulated is the Fundamental Petroleum Policy for Ghana, which requires all players in the oil industry to recognize that "achieving excellence in environmental management, health and safety, and relating well with the community in which a company operates not only contribute to business results by safeguarding people and conserving resources, but also serve as useful indicator of effective management systems".

Additionally, the National Oil Spill Contingency Plan (2010) has been formulated to respond to oil spills of any size in Ghanaian waters. The Plan provides the framework for coordination of an integrated response, definition of responsibilities, reporting and alerting procedures and means of communication, training and exercises, equipment, etc. Plans are far advanced in the preparation of Ghana Petroleum Development Master Plan, which is intended to provide the framework for rational and systematic allocation of resources to address the potential impacts of the development of petroleum resources on the population and the environment. Also being prepared is the National Environmental Policy (NEP), which has the ultimate aim to ensure sound management of the environment and the avoidance of exploitation of resources in ways that may result in irreparable damage to the environment.

Performance-Based Approach

In the performance-based or "selfregulation" approach, which is based on agreements made between government and operators, greater emphasis is placed on setting environmental goals or standards to be met by operators in the industry. This requires the operators to define strategies and plans in order to achieve the overall objectives and criteria set by the regulator. Accordingly, the operators are responsible for providing evidence, assuring that they are complying with the agreements. An example is a legally binding Environment Action Plan (EAP) that is formulated by the operator and subject to reporting and auditing requirements (Technical Meeting Document, 1998). The self-regulation approach focuses on self-inspection (internal audits) by company experts, in consultations with skilled external auditors, in order to check compliance and report to the regulator. It, thus, removes some of the burden of auditing and inspection from government, while allowing the operator flexibility in choosing practical measures to meet the environmental objectives (Technical Meeting Document, 1998). This approach could, therefore, be adopted by the operators in the industry, who will be presented with the opportunity to find other ways of meeting the goals or targets set by government. Thus, the oil companies could be mandated by government to develop Environmental Management Plan (EMP) or Environmental Management System (EMS) to ensure that they operate within the environmental standards for the industry. EMS is a tool which involves continual cycle of planning, implementing, reviewing and improving the processes and actions that will effectively and efficiently enable an organization meet its business and environmental goals (Five Wind International, 2004). This means that there is a review of the system after each cycle to identify areas for further improvement to meet the national environmental standards for the industry.

The EMS, if well implemented, offers a lot of benefits including improved environmental performance, enhanced compliance, pollution prevention, reduction in emissions, resource conservation and reduction in environmental pollution. As part of operational measures, oil companies should develop innovative environmental technologies to be employed in their operations, and develop a proper disposal of generated solid waste. The two types of approach could be achieved through the collaborative efforts of the Ministry of Environment, Science and Technology, the Environmental Protection Agency (EPA), the Ghana Standards Authority (GSA), the oil and gas companies, and other stakeholders in the industry.

Recommended Administrative and Institutional Support. A perfect blend of both prescriptive and performance-based approaches could serve a good purpose in pursuing environmental management in the oil and gas industry. In many countries, performance-based approaches are increasingly being adopted to complement existing prescriptive regulations. Classical examples exist in Norway, the Netherlands and Australia, where the offshore oil industry has been moving to a regime based on goal-setting approach, supplemented by the prescriptive system of regulation (Technical Meeting Document, 1998). However, the mere prescription of environmental codes and setting of standards, as well as the development of EMS, cannot provide the much needed panacea for pollution emanating from the oil and gas industry. Guided by the drawbacks encountered in Ghana's mining industry, it is important that an improved and sustainable strategy be put in place to ensure that oil companies strictly adhere to regulations guiding their activities in the industry, and are not spared any documented punishment if they violate any of the legislations. Accordingly, it is recommended that the following be considered by government in its quest to safeguard the ecosystem whilst exploiting the oil and gas resources: 1. Government should ensure strict control and enforcement of environmental policies; 2. Strengthening existing regulatory framework for environmental protection; 3. Regular and effective monitoring of oil development activities; 4. Periodic update of environmental guidelines; 5. Periodic upward review of fines/penalties to deter potential polluters; 6. Periodic review of the effectiveness of local environmental agencies; 7. Availability of resources for staff development in the regulatory and enforcing agencies; 8. Improved remuneration to prevent violations of legislation by companies and discourage bribery; 9. Tax and duty exemptions on the importation of technologies related to environmental control to encourage firms in both industries to transfer pollution control technology to their establishments; 10. Regular inspection and maintenance of oil installations.

In addition, Ghanaians should be equipped with the necessary knowledge, skills, attitude and motivation for the prevention of pollution and resource deterioration. Furthermore, establishment of conservation pressure groups, with requisite expertise should be encouraged to serve as an appropriate watch dog, providing public education and making sure that the environment is conserved. Environmental education, both formal and informal, should be embarked on to inculcate environmental values and the habits of preservation and conservation among the entire citizenry.

CONCLUSIONS

The exploration, development and production of oil and gas in the Jubilee Oil Field could be associated with ecological degradation, but these effects can be minimized if the Government of Ghana takes steps to develop petroleum industry specific environmental protection guidelines and appropriate regulatory infrastructure including monitoring equipments, compliance enforcement networks and also a deterrent sanction regime. This should employ an integrated approach, involving both prescriptive and performance-based approaches in managing the environment. Thus, the sustainable development of the Jubilee Oil Field has the potential to bring a positive change to Ghana through the preservation of the marine environment and ecosystems, and improvement of the welfare of communities to be impacted by the oil and gas industry, while enhancing the economic prosperity of the nation.

ACKNOWLEDGEMENT

The authors are grateful to Samuel B. Dampare for discussions, and an anonymous reviewer for reviewing the manuscript with valuable comments.

REFERENCES

1. Adjaye R. (2009). Ghana's oil find: technological challenges for upstream skills development. National oil and gas conference on the theme positioning the transport sector for the successful exploitation of Ghana's oil and gas. Paper presented at the Accra International Conference Centre. 15–16 July, 2009.
2. Amnesty International (2009). Nigeria: Petroleum, Pollution and Poverty in the Niger Delta. Amnesty International Publications. Index: AFR 44/017/2009. FRP143. 143 pp.
3. Asafu-Adjaye N. B. (2011). Overview of market opportunity, Ghana. Presentation by Ghana National Petroleum Corporation at the Offshore Europe 2011, Aberdeen, Scotland. 6–8, September.
4. Armah A. K., Biney C., Dahl S. Ø. and Povlsen E. (2004). Environmental

sensitivity map for coastal areas of Ghana. EPA/UNDP Report, Vol. 2.
5. Ayoade M. A. (2002). Disused offshore installation and pipelines: towards sustainable decommissioning. Kluwer Law International. ISBN- 1 3 :9789041117397. pp. 74–77.
6. Barnard R. and Newby T. (2009). Sustainability of terrestrial ecosystem. National State of the Environment Report - South Africa . http://www.ext.grida.no/soesa/nsoer/issues/land/in dex.htm . (Accessed: 7/12/2009)
7. Boely T. and Frèon P. (1980).Coastal pelagic resources. In The fish resources of the Eastern Central Atlantic. Part one: The resources of the Gulf of Guinea from Angola to Mauritania. (J.-P. Troadec and S. Garcia, eds). FAO Fish. Tech. Paper. (186.1):166 pp.
8. British Broadcasting Corporation (1989). Exxon Valdez c r e a t e s o i l s l i c k d i s a s t e r . http://news.bbc.co.uk/onthisday/hi/dates/stories/march/24/newsid_4231000/4231971.stm (Accessed: 18/10/2011)
9. British Broadcasting Corporation (2000). Oil spill damage worsens. http://news.bbc.co.uk/2/ hi/europe/592378.stm (Accessed: 18/10/2011)
10. British Broadcasting Corporation (2011). BP oil spill: the environmental impact one year on. http://www.bbc.co.uk/news/science-environment- 13123036 (Accessed: 18/10/2011)
11. British Petroleum (BP) (2008). Statistical review of world energy. www.bp.com (Accessed: 20/05/2010) Think Ghana (2007). Jubilee Field to flare gas contrary to assurances from Government. http://business.thinkghana.com/pages/industry/20 1012/51461.php (Accessed: 28/9/2011)
12. Darko-Mensah K. O. (2009). Ghana's oil and gas: the people's connection, Daily Graphic, 19 August, 2009.
13. E&P Forum/UNEP (1997). Environmental management in oil and gas exploration and production. An overview of issues and management approaches. Joint E&P Forum/UNEP Technical Publication 37. Oxford. UK.
14. EDM/GNPC (2009). Natural gas transportation and processing project. Environmental and social management framework. Terms of reference. EDM/GNPC Report.
15. IMF (2005). Emergence of the Gulf of Guinea in the global economy.

IMF Working Paper: WP/05/235.

16. EPA (2010). National oil spill contingency plan. Ghana's "National contingency plan to combat pollution by oil and other noxious and hazardous substances". Environmental Protection Agency (Ghana) Report. 103 pp.

17. ERA/CJP (2005). Gas flaring in Nigeria: A human rights, environmental and economic monstrosity. ERA/CJP Technical Report, Amsterdam.

18. Exploration & Production Forum (1994). Atmosphere emission from the offshore oil and gas industry in Western Europe. E&P Forum, December, 1994.

19. Fernández A., Edwards J. F., Rodríguez F., Espinosa de los Monteros A., Herráez P., Castro P., Jaber J. R., Martín V. and Arbelo M. (2005). 'Gas and fat embolic syndrome' involving a mass stranding of beaked whales (Family Ziphiidae) exposed to anthropogenic sonar signals. Vet. Pathol. 42: 446-57.

20. Five Winds International (2004). Environmental management systems. A guidebook for improving energy and environmental performance in local government. Municipal EMS Guidebook. 245 pp.

21. Gao Z. (1998). Environmental regulation of oil and gas industry. Kluwer Law International. London.

22. Ghana News Agency (GNA) (2009). Cyanide spillage, Newmont was negligent. http://www.ghananewsagency.org/s_social/r_869 8/ (Accessed: 2/5/2010)

23. Ghana Oil Watch. http://ghanaoilwatch.org/ index.php?option=com_content&view=article&i d=1733:gas-flaring-imminent-jubilee-field&catid=6:ghana-oil-a-gas-news&Itemid=27 (Accessed: 28/9/2011).

24. GNPC (2009). Exploration and production history of Ghana. http://www.gnpcghana.com.

25. Harremoës P. (ed.) (1998). Water quality processes. Department of Environmental Engineering, Technical University of Denmark, Lyngby, Denmark.

26. Kloff S. and Wicks C. (2004). Environmental management of offshore oil development and maritime oil transport. A background document for stakeholders of the West African Marine Eco Region. IUCN working paper.

27. Kosich D. (2010). Ahafo mine cyanide spill prompts NGOs to question international cyanide code validity. http://mineweb.com/mineweb/view/mineweb/en/page72068?oid=96302&sn=Detail (Accessed: 30/01/2010)
28. Mañe D. O. (2005). Emergence of the Gulf of Guinea in the global economy: Prospects and challenges. An IMF Working Paper. WP/05/235.
29. MFRD (2003). 1986-1995 Report on the Fisheries Research and Utilisation Branch of the Fisheries Department. (Marine Fisheries Research Division Directorate).
30. McCauley R. D. (1994). Environmental implications of offshore oil and gas development in Australia - seismic surveys. Australian Institute of Marine Science, Townsville, Queensland. 121 pp.
31. Myjoyonline (2010). Newmont pays GH¢7 million in spillage compensation. http://news.myjoyonline.com/business/201004/44661.asp (Accessed: 3/05/2010)
32. NCF/WWF/IUCN (2006). Niger Delta Natural resources damage assessment and restoration project scoping report, May 2006. Jointly prepared by Nigerian Conservation Foundation (NCF), World Wildlife Federation (WWF), UK and International Union for Conservation of Nature (IUCN), Commission on Environmental, Economic and Social Policy, with Federal Ministry of Environment (Abuja).
33. Patin S. (1999). Environmental impact of the offshore oil and gas industry. EcoMonitor Publishing. East Northport, N. Y. 425 pp.
34. Popper A. N. (2003). The effects of anthropogenic sounds on fishes. Fisheries, 28(10): 24–31.
35. Pyagbara L. S. (2007). The adverse impacts of oil pollution on the environment and wellbeing of a local indigenous community: the experience of the Ogoni people of Nigeria. United Nations Department of Economic and Social Affairs Report. PFII/2007/WS.3/6.
36. Reuters (2010). Factbox: Gulf oil spill impacts fisheries, wildlife, tourism. http://www.reuters.com/article/2010/05/30/us-oil-rig-impact-factbox-dUSTRE64T23R20100530 (Accessed: 18/10/2011)
37. Sadiq R., Veitch B., Williams C., Pennell V., Niu H., Worakanok B.,

Hawboldt K., Husain T., Bose N., Mukhtasor and Cole C. (2002). An integrated approach to environmental decision-making for offshore oil and gas operations. Canada-Brazil Oil & Gas HSE Seminar and Workshop, March 11–12.

38. Sala-i-Martin X. and Subramanian A. (2003). Addressing the natural resources curse: an illustration from Nigeria. IMF Working Paper. WP/03/139.

39. Steiner R. (2008). Double standards? International best practice standards to prevent and control pipeline oil spills, compared with Shell practices in Nigeria. Friends of the Earth, Netherlands. 61 pp. November, 2008.

40. Sunu-Attah V. K. (2009). Ghana's oil and gas sector strategy: an overview of the scope and development plans for the sector-GNPC. National oil and gas conference on the theme "Positioning the transport sector for the successful exploitation of Ghana's oil and gas" Accra International Conference Centre. 15–16 July, 2009.

41. Technical Meeting Document (1998). Environmental practices in offshore oil and gas activities. A compilation of information discussed in the working sessions at the international expert meeting in Noordwijk, The Netherlands. 17–20 November. 91 pp.

42. Tyack P. L. and Miller E. H. (2002). Vocal anatomy, acoustic communication and echolocation. In Marine Mammal Biology. (A. R. Hoelzel, ed.), pp. 142–184. Blackwell Science Ltd., Oxford, UK.

43. UNEP (2002). Africa environment outlook: past, present and future perspectives. United Nations Environmental Programme Report. 400 pp.

44. UNEP (2011). Environmental Assessment of Ogoniland. United Nations Environmental Programme Technical Report. 262 pp.

45. WHO/EC (2002). Eutrophication and health. Local authorities, health and environment. World Health Organization/European Commission Report. Briefing Pamphlet Series 40.

46. World Bank (1995). Defining an environmental development strategy for the Niger Delta, Vol. 2. Industry and Energy Operations Division West Central Africa Department. 35pp.

47. World Energy Outlook (2009). www.worldener gyoutlook.org.

(Accessed: 31/07/ 2010).

CITAION

P. A. Sakyi, J. K. Efavi, D. Atta-Peters, and R. Asare, Ghana's Quest for Oil and Gas: Ecological Risks and Management Frameworks, doi:10.1016/j.lithos.2013.08.010.

Chapter 7

Hydrochemical and Hydrogeological Impact of Hydraulic Fracturing in the Karoo, South Africa

G. Steyl[1,2] and G. J. van Tonder[3]

[1]Golder Associates Pty Ltd, Milton, Queensland, Australia
[2]Department of Chemistry, University of the Free State, South Africa
[3]Institute for Groundwater, University of the Free State, South Africa

ABSTRACT

Hydraulic fracturing has become a prevalent public and regulatory issue in most countries developing shale gas. South Africa has only recently been exposed to terrestrial gas resource development and this has created unique regulatory issues which are currently being resolved. One of the key issues under debate is the protection of groundwater resources in rural areas, since most of South Africa's rural and some inland cities are dependent on groundwater for potable water supply. A second concern is the infrastructure requirements to handle the material movement processes during the development of each wellfield and subsequent processing of waste generated on site. Regarding the waste material production, a phased approach is required which considers the initial well development activities, production and subsequent well abandonment. Each phase has a unique risk associated with it and thus would require different management options. At the current stage most of the focus is on the initial stages of well development but the long term view has been neglected to some extent. Due to the unique geological structure of the Karoo, the presence of dolerite structures, a number of risk mitigation methods might be required to succesfully develop hydraulically fractured wells. In all aspects the chemical and hydrogeological impacts related to wellfield development cannot be ignored in the Karoo aquifer system, as it may directly influence human and environmental health. This paper will present chemical perspective on the hydraulic fracturing perspective that will deal with the impact of hydraulic fracturing fluid and flowback water. Additionally, the interaction of wellfield development and hydrogeology of the Karoo area will be discussed and how it relates to future water quality issues.

INTRODUCTION

This paper attempts to summarize the current knowledge on hydraulic fracturing and production issues related to shale gas in South Africa. The observation and findings made in this work is neither totally comprehensive nor exhaustive since no data is available in the public domain on hydraulic fracturing and it associated activities in South Africa. The exploration for natural gas resources in terrestrial South Africa has been conducted since the mid-1960's, however no exploitable source could be located. Limited gas was however found in the tight shale formations of the Ecca Group at an approximate depth of 2000-4000 metres below surface. The potential current shale gas reserve in the Karoo shales is estimated to be 485 trillion cubic feet, which would make it the fifth largest shale gas field in the world [1]. In geological terms the Karoo Supergroup refers to an extensive geological sequence (100-260 million years old) which consists of sedimentary and igneous rocks. Most of the Karoo Supergroup is located in South Africa and the Great Karoo has an area of more than 600 000 km^2.

Due to present energy shortfall in South Africa, the requirement for new energy sources have gained new momentum and part of this new focus is on shale gas in Karoo type formations. The most interesting aspect of this is that the area available for natural gas development is substantially larger than just the Karoo, with exploration areas covering six of the nine provinces in South Africa [2]. The development of shale gas resources was initiated in late 2009 but were halted due to a moratorium in early 2011. This has subsequently been lifted in September 2012. There are currently five pending applications related to exploration in the Karoo (Figure 1), three belong to Shell and one each to Falcon Oil and Gas and Bundu Gas and Oil Exploration [3]. To the north is located the petrochemical group Sasol gas exploration area, however plans have been put on hold by the company until further notice. An exploration area has also been awarded to Anglo American adjacent to the Sasol area (Figure 1).

Hydraulic fracturing was developed in the United States of America in the late 1940's to assist in the stimulation of oil and natural gas wells [4-6]. The number of wells that incorporates hydraulic fracturing increases by the day since oil and gas production is increased by this technique [5].

Development of Shale Gas in South Africa

The Shale Gas project aims to target the carbonaceous shales of the Ecca and Dwyka Groups, but the stratigraphic units in question vary in lithological makeup along strike as one proceeds from the Cape to the Free State/Natal (KZN) regions. The initial reasoning was to only target zones of the Whitehill formation of the lower Ecca, which is a carbonaceous shale unit characterized by its distinctive white weathering in outcrop. The distribution of the Whitehill Formation with its marine setting led to further investigation into the dynamics of the Main Karoo Basin and other stratigraphic units equivalent to the Whitehill to extend the potential target areas [2]. A revised set of source rocks were identified with the main target zone now being identified as carbonaceous shales of the Lower and Upper Ecca Group with subordinate interest in the Dwyka Shales (Figure 1). The source formations have been extended to include the following:

- Whitehill Formation (Cape region)

- Prince Albert Formation (Cape region)
- Volksrust Formation (Free State and KZN regions)
- Vryheid Formation (Free State and KZN regions)
- Pietermaritzburg Formation (Natal region)
- Dwyka Shales (All regions, where shallow enough).

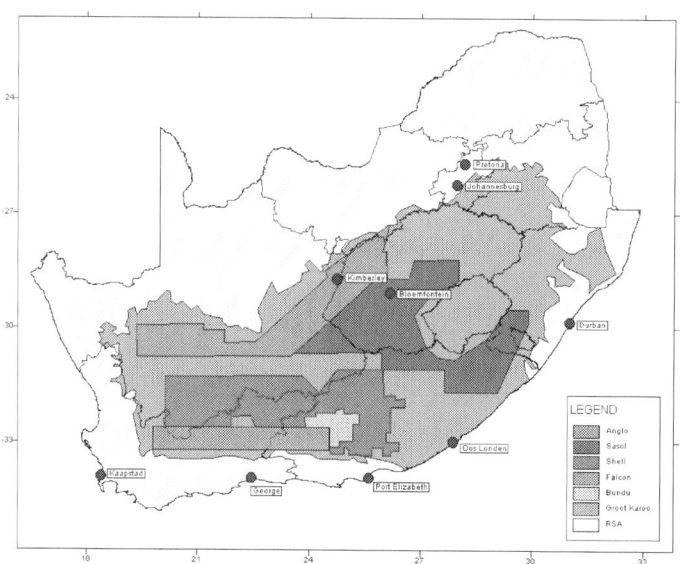

Figure 1: Regional map of South Africa, showing the exploration rights and companies associated with these permits [2].

The research into the Shale Gas deposits of the USA led to a revised set of geochemical and petrophysical parameters that are based on the criteria set by Jarvie [2] to include the following:

- Total Organic Carbon and its composition (dead carbon, free gas, etc.) → 1% or more.
- Kerogen Type → Determines hydrocarbon types as well as adsorption/desorption properties.
- Vitrinite reflectance and Tmax (maximum temperatures that rocks were subjected to during hydrocarbon production) → Thermal Maturity with reflectance values of 1.35-2.5, Tmax values vary and can be very high (between 400°C and 580°C as seen in the Barnett Shale).

- Rock Eval Hydrogen Index <100.
- Porosities and other physical properties related to gas flow.
- Calculations of hydrocarbon generation, expulsion and retention.

From a South African perspective, the Rowsell and De Swardt's study [7] of the maturation indices pertaining to the Karoo Basin can be used to identify areas prospective for gas generation:

- Temperature Range → ±130°C to 170°/180°C.
- Vitrinite Reflectance of 1.35-2.5.
- CR/CT Ratio of about 0.85 to 0.94.
- Total Organic Carbon and its composition (dead carbon, free gas, etc.) → 1% or more.

Geology and Gas Plays in South Africa

In South Africa, shales containing significant organic carbon are restricted to the Ecca Group of the main Karoo Basin, smaller basins in the northern part of South Africa and to the Bokkeveld Group in the southernmost part of South Africa [7]. These muds became buried and lithified over tens to hundreds of millions of years and generated various hydrocarbons with increasing depth of burial and increasing temperature (Figure 2). Between 2-4 km burial depth, oil is produced, between 4-5 km, wet gas is produced and between 5-6 km, dry gas, including methane, is produced. Deeper burial results in low-grade metamorphism, the termination of hydrocarbon generation and the formation of graphite from the organic material. In South Africa, shales of the Bokkeveld Group have undergone low-grade metamorphism and no longer have a capacity for hydrocarbon generation. However, after comprehensive investigations it was confirmed that Ecca Group shales might have the potential to generate dry gas south of the 29°S [7]. Further north, the shales have been less deeply buried and have a potential for oil generation except where younger igneous dolerite intrusions have locally increased the thermal maturity leading to the generation of dry gas [7].

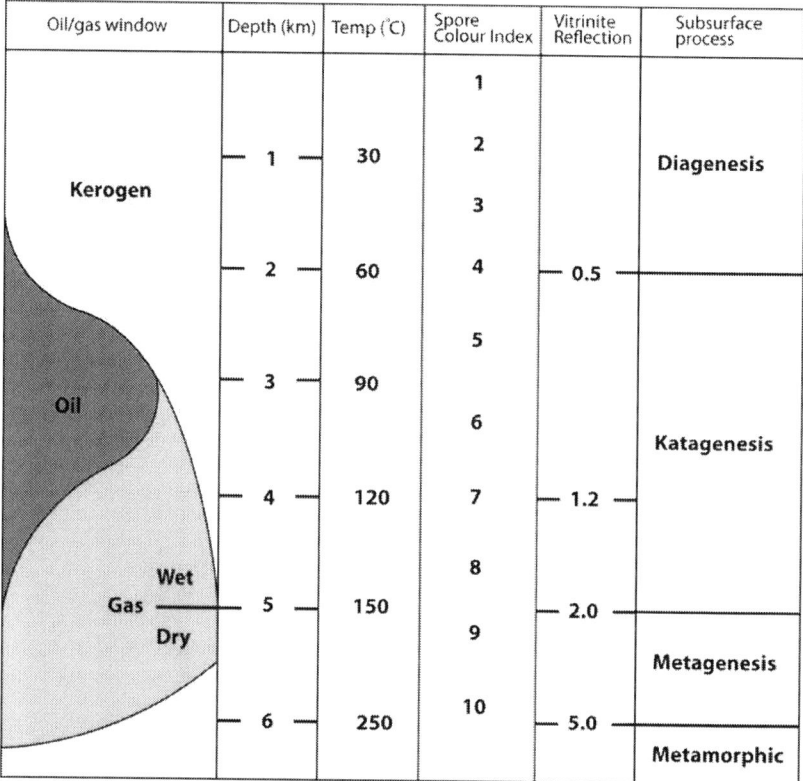

Figure 2: Hydrocarbon generation and thermal maturation indices and maturation stages plotted against depth of burial [8].

Total organic carbon within the shale is an important parameter, since there is a linear relationship between total organic carbon and gas content, as in the Barnett Shale in the Fort Worth Basin of Texas [9]. Thickness is also important, as most of the gas produced is from areas where the shale is between 90 and 183 metres thick [9]. However, more recently, it has become technically possible to produce gas from shale units as thin as 10 to 15 metres [10]. Within the main Karoo Basin (Figure 3), there are reports of natural gas occurrences both at surface and at intervals in the deep wells drilled by Soekor between 1965 and 1977. Furthermore, varying quantities of gas were obtained by desorbed gas analysis undertaken by Soekor on Ecca Group shale samples retrieved from the deep well cores [7].

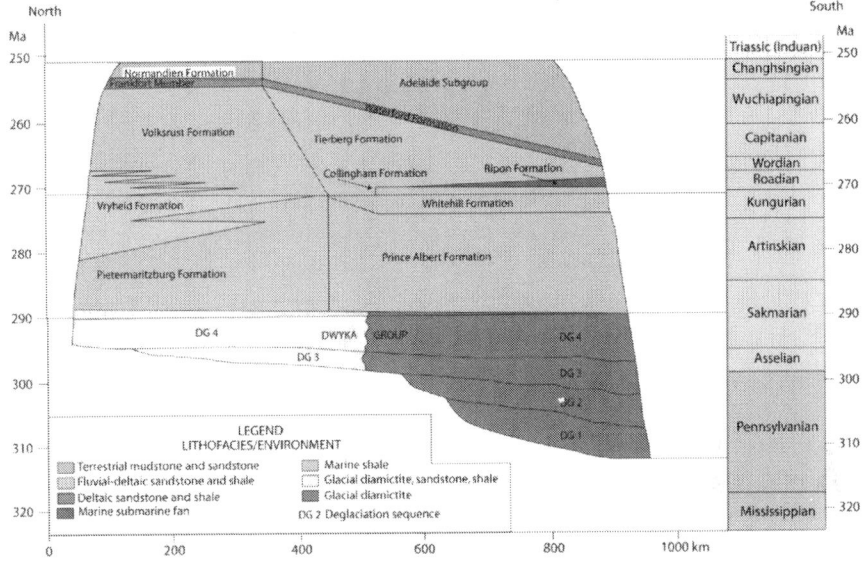

Figure 3: Distribution in time and space of the Dwyka Group, Ecca Group and Adelaide Subgroup in the main Karoo Basin, South Africa, showing lithofacies, environment and stratigraphic relationships. Modified from Fig. 7 of Veevers et al. [11]. The geologic timescale is from Gradstein et al. [12].

It was found that only the lower Ecca Group shales [7] within the dry gas window south of latitude 29°S have comparable total organic carbon contents to those of producing shales elsewhere in the USA (Table 1). The upper Ecca Group shales, namely the Tierberg Formation [13], average only 1.2 percent organic carbon [14], which is significantly lower than the 3 to 12 percent range applicable to producing shales (Table 1). The Dwyka Group also contains black shales with between 0.1 and 4.3 percent total organic carbon, averaging 1.9 percent [15; 14]. However, these shales are thin and restricted, being interbedded with diamictite and sandstone, with the thickest shales (50 to 60 m) occurring in only 3 out of 45 deep wells investigated. The lower Ecca Group comprises black, organic-rich shale of the Whitehill Formation [16] overlying dark grey shale of the Prince Albert Formation (Figure 3) [17]. The Whitehill Formation pinches out northeastwards along a line stretching from Hertzogville in the Free State to Coffee Bay in Eastern Cape Province [14]. Northeast of this line, the Whitehill Formation correlates with the middle part of the sandstone-dominated Vryheid

Formation and the Prince Albert Formation grades into shale of the Pietermaritzburg Formation (Figure 3). In the area between Coffee Bay and Harding, the Whitehill and Vryheid Formations are separated by a continuous shale succession [18].

The gas production probability was delineated by Rowsell and De Swardt [7] using the results of desorbed gas analysis on core samples from the deep Soekor wells. The gaseous hydrocarbons (methane to pentane) are absorbed on to the fine-grained constituents of shales and can be desorbed by low-temperature acid hydrolysis [7]. Samples yielding high proportions of C1 gas (methane) and C2/C1 (Ethane/Methane) relative to C3/C1 (Propane/Methane) indicate a potential for dry gas. The trend of increasing maturity due to increasing depth of burial southwards across the basin is supported by the results from other parameters, namely vitrinite reflectance, CR/CT ratios, illite crystallinity and spore colour index. For dry gas generation, vitrinite reflectance values should be between 2 and 5 percent. In the main Karoo Basin south of latitude 29°S, values for shale of the Ecca and Dwyka Groups vary between 1.8 and 4.4 percent [7]. Branch et al. [19] measured vitrinite reflectance values between 3.5 and 5.3 percent for shale of the Whitehill Formation and between 4.0 and 6.4 percent for shale of the Prince Albert Formation in well SA1/66 in the southwestern part of the basin some 60 km north of the basin margin. These correspond to the dry gas and metamorphic maturation stages, which indicates that shales in the southern extremity of the present basin are over-mature and can no longer generate dry gas. CR/CT ratios (residual, non-volatile carbon after pyrolysis to total carbon in the kerogen or organic material) gives an indication of the ability of the shale to produce additional amounts of hydrocarbons if heated to sufficiently high temperatures with lower ratios corresponding to higher potential. The results more or less correspond to the findings of the desorbed gas analysis [7]. Illite crystallinity or Kübler index is a measure of the width in millimetres of the 10 Å diffraction peak at half its height. It gives an indication of the maturity level of the shale with decreasing indices corresponding to increasing maturity [7]. The results of Soekor's investigations indicate a trend of increasing Kübler index from south to north across the Karoo Basin in shales of the Ecca and Dwyka Groups. In the southern part of the basin, south of approximately 30°S, the average indices are less than 4 and correspond to the metagenesis stage and possible preservation of dry gas. Comparative data to the Marcellus Shale and Barnett Shale is presented in Table 1. The percentage organic carbon detected in these shale formations in the USA are similar to those determined for the Whitehill, Prince Albert and Pietermaritzburg formations. Additionally, the thickness of the formations are also comparable to the Marcellus and Barnett shales. However, the Tierberg Formation and Volksrust Formation can also be possible future targets for shale gas exploration since these formations are considerably thicker than the USA counterparts but at a lower organic carbon content.

Table 1: Comparative results of estimated percentage organic carbon and thickness of formation [2]

Unit or Formation	Percentage organic carbon (%)	Thickness (metres)
Marcellus Shale	0.3-20.0	12-270
Barnett Shale	0.5-13.0	15-300
Karoo Basin-Whitehill Formation	0.5-14.7	0.4-72
Karoo Basin-Prince Albert Formation	0.3-12.4	30-500
Karoo Basin-Pietermaritzburg Formation	0.3-11.6	0.8-420
Karoo Basin-Tierberg Formation	0.3-5.2	400-1300
Karoo Basin-Volksrust Formation	0.3-5.9	250-415
Karoo Basin-Dwyka Group	0.1-4.1	0-58

In the following sections a comparative analysis of known attributes of the Karoo shales and Marcellus Shale will be further developed as well as the impact on the hydrogeology and hydrochemical components.

PROBLEM FORMULATION

The major concern to date in the Karoo is the contamination of readily accessible water supply, i.e. surface water or groundwater resources. The development of an unconventional gas field does not occur in a matter of months, with a typical initiation phase of 10 years before gas production can continuously take place [20; 3]. In the instance of South Africa, a number of issues restrict the development of an effective gas extraction project. The infrastructure for gas transport (pipelines) in South Africa is very limited since no conventional terrestrial gas fields exist within the borders of the country. The Soekor wells drilled between 1965 and 1977 have yielded only tentative clues to the availability of gas in the Karoo basin. In this regard the major gas companies have to do a comprehensive exploration and verification program that could last from 3-6 years depending on the geological complexity of the development area. This would be followed by a pilot study to evaluate the basic characteristics of the reservoir which can be done on a number of sites simultaniously over a period of 2-4 years. Finally, if the gas in place is adequate then the process can be developed into full production of gas which can last for 30-100 years depending on gas prices and availability within the shales.

The geology of South Africa is quite varied considering the land size. One distinguishing feature of the geology is the presence of dolerite sills and dykes (Figure 4). The stack of sedimentary strata above the targeted formation in the Karoo consists of a succession of shale, mudrock, sandstone and dolerite. Each of these rock-types are generally characterised by low matrix transmissivities (between 0.5-50 m^2/day) [21]. These values were obtained from pump tests carried out on Karoo aquifers less than 200 m deep. Matrix transmissivities at greater depth would therefore be expected to be even less than these values, however this still needs to be confirmed in the future. Dolerite matrix has also been found to be quite impermeable [22] but due to the process of intrussion it can also act as an conduit. It is expected that the process of well field development would take into consideration the presence of these structures and that upward injection and production fluids would be limited.

Many of the areas where the shale formations have the potential to represent a good prospective target for exploration are also characterised by multiple dolerite intrusions. Drilling in a dolerite sill environment will face challenges that can be overcome if sufficient investigation is carried out on these intrusive structures at depth. There is sparse information on the structure of deep dolerite sills and associated deep groundwater and water strikes in the Karoo lithostratigraphic formations. All available data comes from groundwater exploration drilling at shallow to medium depth (< 300 m). Several groundwater strikes were intercepted at that depth [2]. Below this depth, the presence of deep water strikes in the Karoo formations and associated dolerite, their yields and the composition of the water are still a matter of debate.

Figure 4: A regional map showing a subsection of the Karoo Supergroup with blue patterns indicating sills while green to red represents dykes in the area [2; 23].

The key question is: Can dykes act as vertical conduits for groundwater flow or hydraulic fracturing fluids? From current literature available [2; 23], it is clear that many water strikes occur between 0-70 metres below ground level (i.e. in the weathered zone) and are found at the contact dyke-sediment. Below 70 metres the water strikes are found along transgressive fractures. The main mechanism of flow dynamics at depth and around dykes is associated with sub-horizontal fractures. These fractures are not linked to one another and collect water from the matrix (dual porosity medium). The influence of dolerite dykes on vertical groundwater circulation at depth seems therefore to be limited, but cannot be excluded due to limited data availability. The T-values from different case studies also show that permeability of the dykes is too low to allow for major flow in the dykes themselves [24].

However there is evidence of a natural connection between deep groundwater systems and the surface, as evidenced by sixteen naturally occurring warm water (thermal) springs (26-41°C) in the main Karoo Basin south of latitude 28 degrees [25; 26]. These waters originate at a maximum depth of between 450 m and 1 150 m, as calculated from the geothermal gradient and the surface temperature of the waters. All the waters are according to Kent [25] are originally meteoritic and mainly deviate in composition due to differences in different compositions of the different rock lithologies associated with the spring, indicating also the presence of connate water. The waters of the central and eastern Karoo have NaCl as the prominent constituent with total dissolved solids ranging from 480-780 mg/l. A few springs are, however characterised by high $NaHCO_3$ and SO_4 contents, e.g. Stinkfontein, south of Beaufort West and the spring at Cradock. Biogenic methane is one of the main gases commonly associated with the hot springs in the main Karoo Basin and in some instances constitutes the only gas present. The other gases present are mainly H_2, N_2, He and Ar [25].

Currently within popular literature hydraulic fracturing and well field development is grouped together under the phrase of fracking. This is clearly incorrect as gas companies refer to fracking as the process of hydraulic fracturing of the formation (shale or gas containing strata) for the purpose of increasing the porosity and permeability of the system to extract shale gas. In order to effectively evaluate the risks associated with hydraulic fracturing, it is best to assess the whole process in which an unconventional shale gas well is developed. This will include the drilling, hydraulic fracturing, well completion, production phase and post closure of the well field itself.

RESULTS AND DISCUSSION

Due to the lack of current data on the Karoo (Permian), secondary sources are required to infer possible issues in this area. Firstly, to assist in the investigation international studies were required for a comparative basis to describe the influence of shale gas development programs. These areas included the Marcellus (Devonian, Pennsylvania), Antrim (Upper Devonian, Biogenic, Ohio) and Barnett (Mississippian, Texas) shale plays. These were selected due to their state of unconventional gas development and regulatory framework. One report that has recently been made available for public scrituny that contains some measurement data has indicated some interesting trends [27]. The report summarises both sampling from vertical and horizontal drilled wells and reports a full range of chemical and flow data. In addition to this report it was also required to evaluate the hydraulic fracturing fluid composition used in the stimulation of the shale gas well. Since it is uncertain what specific set of chemicals will be used in the hydraulic fracturing event, it was deemed the best possible solution to assess the generalised composition of these fluids. In regard to hydraulic fracturing process it should be kept in mind that although it is referred to as a single process it consists of multiple steps. Each step has a purpose in the hydraulic fracturing event as well as the transport of the propanant down the hole.

Hydraulic Fracturing Process

Considering the chemicals used during the hydraulic fracturing process, recent publication of hydraulic fracturing fluid compositions has significantly increased the transparency in the use of these chemicals [28; 29]. However, when examining the reported values in the component information disclosure, some reports indicate that there is still some components that are not listed and are most likely proprietary [30]. In the current paper only a single hydraulic fracturing composition is considered, i.e. gel hydraulic fracturing fluid. A number of hydraulic fracturing fluid setups does exist which can either be based on water (slick water), gel, hybrid, foam or gas (air, inert or petroleum gas). The type of hydraulic fracturing fluid used is dependent on a number of factors and service company preference [28].

A recent investigation by the House of Representatives in the USA [31] found that a list of 750 chemical compounds were used from 2005 to 2009. A number of chemical compounds that have been reported, included 29 chemicals that are known or possible human carcinogens and are regulated under the Safe Drinking Water Act or listed as hazardous air pollutants under the Clean Air Act [31]. BTEX compounds–benzene, toluene, xylene, and ethylbenzene–appeared in 60 of the hydraulic fracturing products used between 2005 and 2009. The hydraulic fracturing companies injected 43.1 million litres of products containing at least one BTEX chemical over the five year period. In many instances, the oil and gas service companies were unable to provide the Committee with a complete chemical makeup of the hydraulic fracturing fluids used [31]. Between 2005 and 2009, the companies used 355 million litres of 279 products that contained at least one chemical or component that the manufacturers deemed proprietary or a trade secret [31]. The practice of using BTEX is currently being phased out due to known issues [31].

Interestingly, most of the chemical components of hydraulic fracturing fluids can be described as either LNAPLs and DNAPLs from a South African context. In addition regarding the interpretation of the Water Act of South Africa, an unwanted consequence may result from the process of hydraulic fracturing. Most notably the process increases the permeability and hydraulic conductivity of the zone that is fractured. This in part can constitute an aquifer at a substantial depth from surface, in this instance it would be regarded as a controlled activity with a host of requirements that needs to be addressed to satisify regulatory practice.

By means of an illustrative example it is possible to get a rough estimate of the extent of chemical usage in hydraulic fracturing. It has been stated that a vertical hydraulic fracturing process requires 1×10^6 litres of fluid; in contrast a single horizontal hydraulic fracturing process requires 10×10^6 litres of fluid. A pamphlet recently released by Energy in Depth gave a generic summary which stated the percentage composition of hydraulic fracturing fluid as reported by the Department of Energy [32]. If these values are taken as a lower limit then the following deductions can be made from Figure 5. Water and sand component of the hydraulic fracturing process constitutes 99.51% of the total volume used.

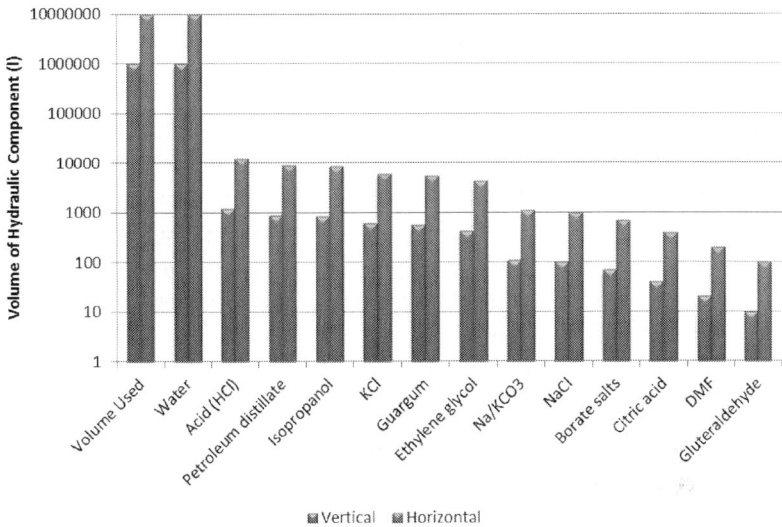

Figure 5: Generalised volume of hydraulic fracturing component used in well stimulation. Vertical well and horizontal well is indicated in blue and red bars, respectively.

Additives employed in the vertical or horizontal fracturing is present in scales approaching tonnes. Chemicals that are of special concern in large quantities are the acid phase, petroleum distillate and isopropanol. The acid phase is composed of hydrochloric acid (10-15%) and is usually part of the first phase of fluids to be injected into the well. The main aim of the acid phase is for cleaning the perforations and initiating fissures in the near-wellbore rock (acid-etching). A secondary consequence is that the acid injected does interact with the host rock formation which can mobilise certain metals, but the mobilisation is dependent on acid concentration and exposure to host rock formation [33]. Petroleum distillates and isopropanol is listed chemicals of concern (carcinogens, SDWA regulated chemicals and hazardous air pollutants in the USA) and is still used in hydraulic fracturing activities [31; 28; 30]. Other chemicals that are also classified as chemicals of concern are ethylene glycol, dimethyl formamide (DMF) and hydrochloric acid. If these components are added together more than 3410 and 34100 litres of chemicals of concern is injected into a well to develop a vertical or horizontal hydraulic fractured well, respectively. These values represent a single hydraulic fracturing event and the whole process is repeated if another section is hydraulically fractured in a well. It is important to note that it is assumed that the additives represent 0.49% of the total volume, but it can be as high as 5% in some instances, depending on field circumstances (geology, depth, anisotropic stress, water content and stability).

At this stage the most significant threat that hydraulic fracturing fluid can pose is an uncontrolled spill at surface [34]. This is due to the fact that once the hydraulic fracturing fluid has been injected into the subsurface, it reacts with the specified target components as well as the geological formation and subsurface water it comes into contact with. It is at this stage that the hydraulic fracturing fluid can undergo a number of chemical and physical processes to either precipitate, mobilise, react or undergo physical transformations (adsorption and absorption). In either instance the chemical component has been altered.

However, with current internal practices developed in the gas companies the likelihood of an uncontrolled spill have been significantly reduced. It is generally in the companies own best interest to minimise these events as it can affect future gas development rights and litigation. Spills that do occur on site is usually dealt with immediatly or a remediation plan is put into place [34].

Backflow Event after Hydraulic Fracturing

The current section is focused on a report produced by Hayes [27] for the Marcellus Shale Coaliton. It is one of the few publically available documents that give an indepth report on injected and produced water in a hydraulic fractured well system. The report is used as an illustrative example and it is recognised that the water qualities associated with the Karoo Supergroup will most likely differ. It should be noted that although flowback water is used in this section, that there is no decernable difference between the classification of flowback water and produced water (Table 2). Instead it is an artificial deliniation depending on who has currently control of the site, i.e., hydraulic fracturing team or the production team. Additionally, this section will be used to illustrate the mass of salt produced from these well systems, which in turn would indicate treatment requirements and disposal volumes. It is assumed that the salt will be present as a dry material that would be disposed of in an environmentally approved manner. From a South African perspective, the most likely development of gas well fields will be multiple wells on a single pad. This is due to infrastructure requirements and safety considerations.

The average flowback percentage of vertical and horizontal hydraulic fracturing wells are 43.7% and 25.3%, respectively. In Table 2 the average hydraulic fracturing volume used for vertical and horizontal wells are 5.8 million litres and 13.7 million litres, respectively. This would indicate that more than 50-70% of the fluid injected has been absorbed by the formation. In either instance it does represent a potential source of produced water over time and it is unclear from present data what the potential might be. Factors that could influence the production of water in shales is the current hydrogeological environment of the shale formations, i.e. hydraulic head (pre-hydraulic fracturing), hydraulic conductivity (pre-hydraulic fracturing), porosity and storativity.

In the remainder of this paper the focus will be on the horizontal well systems only and their associated produced volumes and chemical composition. The total dissolved solids for these selected flowback wells have also been included in Table 3. It should be kept in mind that the backflow water not only consists of hydraulic fracturing fluid but also of chemicals that were produced from the geological formation in which the hydraulic fracturing event took place, thus resulting in a mixture of hydraulic fracturing fluid and shale chemical constituents. The volume of water is also a representation of water injected and water present in the shale, which initially depends on the storativity of the shale and the porosity. Most notable of the tables presented here is that there is a number of missing data points, in either the flowback volumes or total dissolved solids concentration values. In some regard this reduces the usefulness of the data but it does give a good indication of expected volumes and salt loading over time.

Table 2: Reported hydraulic fracturing and flowback volumes from Hayes report [27]

Site	Well Type	Hydraulic Fluid (HF) Total Volume (l)	Cumulative Volume of Flowback Water (FW)				%FW/HF
			Day 1* (l)	Day 5 (l)	Day 14 (l)	Day 90 (l)	

A	Vertical	6,366,805	628,000	1,662,371	2,388,466		37.5
B	Vertical	14,979,147	174,091	1,714,201	2,180,988	2,844,283	19.0
C	Horizontal	23,248,077	525,930	1,534,545	2,542,366		10.9
D	Horizontal	3,361,627	453,750	1,284,140	1,580,016	1,778,273	52.9
E	Horizontal	8,505,821	1,360,931	3,232,212	3,912,677	4,082,794	48.0
F	Horizontal	12,400,214	520,206	1,721,832	1,960,472	2,768,446	22.3
G	Horizontal	19,701,865	193,806	1,191,292	1,982,731	2,969,406	15.1
H	Vertical	5,729,107	634,041	2,602,463	3,383,568	5,045,462	88.1
K	Horizontal	11,252,167	914,336	1,274,442	1,506,087		13.4
M	Horizontal	15,770,745	2,610,412	2,851,437	3,135,707		19.9
N	Vertical	1,818,020	386,657	438,646	483,798	562,020	30.9
O	Horizontal	15,375,026	815,764	3,052,874			19.9
Q	Vertical	3,750,987	209,068	568,698	809,245		21.6
S	Vertical	2,616,931	332,919	1,245,189	1,485,736	1,704,821	65.1

Table 3: Concentration of Total Dissolved Solids from Selected Sites (mg/l)

Site	Day 0*	Day 1	Day 5	Day 14	Day 90
C	719	24,700	61,900	110,000	267,000
D	1,410	9,020	40,700		155,000
E	5,910	28,900	55,100	124,000	
F	462	61,200	116,000	157,000	
G	1,920	74,600	125,000	169,000	
K	804	18,600	39,400	3,010	
M	371			228,000	
O	2,670	17,400	125,000	186,000	

The average chemical salt loading in the return water was in excess of a 95 000 mg/l (Figure 6). Considering these values an expected salt load produced from a single well would be in the range of 241 tons of material, which would require adequate disposal regulations since the waste would contain materials classified as harmful to the environment (Sr, Ba, Li, Cl and Br). A further consideration in processing the material would be the quantity of salts produced during a specified time period. Data reported by Hayes [27] were analysed to derive salt loads at reported day intervals at which chemical sample analysis were performed (Table 4). From the data presented the salt loading values vary considerably over production time and that no singular analysis can be used to determine when the most salt from the hydraulic fracturing well would be produced. This is due to different geologies as well as hydrogeological factors (porosity, permeability and water content of the formation). Secondly, salt loads vary from as little as 45 tonnes to 439 tonnes at 90 day, indicating that a significant quantity of salts is produced from each of the respective wells. The cumulative salts produced from these six wells are in the order of 1 920 tonnes which should be disposed of in an environmentally sound methodology.

In order to determine the 90 day values, a linear regression method was used to fit the data to a logaritmic function. Cumulative salt loading values were used since it was composed of both the flowback volume and total dissolved solids (TDS) value. It was assume in the calculations that the decrease in flow volume would continue to follow a logaritmic function, as would typically be expected from a production well. The salt loading (TDS) had a similar pattern and could be expcted to increase in the same methodology for the 90 day time period. If these values are not considered for the 90 day production then the 14 day production in salt loading is expected to be 1 350 tonnes at an average of 169 tonnes.

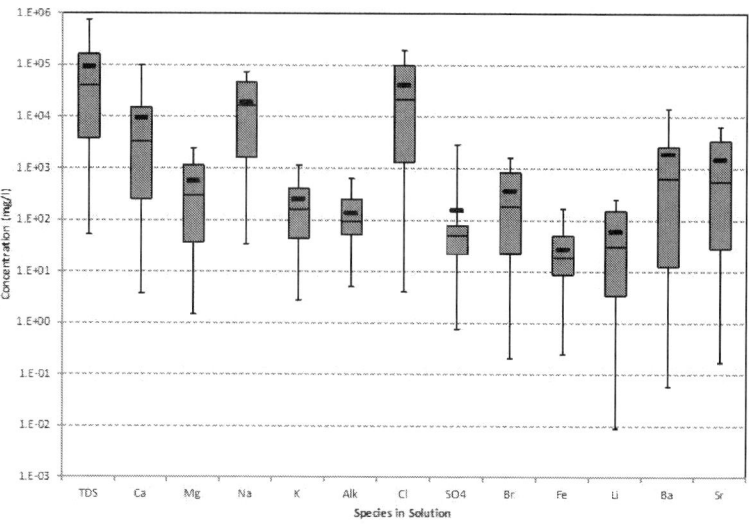

Figure 6: Box-and-Whisker diagram presenting the average distribution of sampled sites chemical components.

Table 4: Cumulative salt loads in tons at a specific day for the respective sites. Values with * indicate projected values.

Cumulative salt load	Day				
Site	0	1	5	14	90
	(tons)	(tons)	(tons)	(tons)	(tons)
C	17	13	75	186	287*
D	5	4	38	65	96
E	50	39	142	227	353*
F	6	32	171	209	349*
G	38	14	139	273	439*
K	9	17	31	32	45*
M	6			65	
O	41	14	294		

Since all of the data which is available from hydraulic fracturing events are based on the Marcellus shale areas in the USA a question arose to the effect as how the Karoo shales compare the Marcellus shale. In order to investigate this question, Whitehill samples were collected from the Geological Department at the University of the Free State and subjected to a leaching test in acid. The results obtained are reported in Table 5 under the heading of Karoo. To draw a comparison between the shales the average chemical analysis of produced water from the Hayes report [27] and average composition of shales [35] were included. Due to different analysis methodologies and production environments these values could not be directly compared, instead ratios of the major elements were used to determine if a possible correlation did exist (Table 6). In general a good correlation existed between the reported sample compositions in the Hem and Karoo data, with all results of the ratios within the same order when compared to each other. In contrast the Hayes report differed notably in the Ba/Ca, Ba/Li and Ba/Mg ratios which could possibly indicate that the use of hydraulic fracturing additives might have changed the chemical character of the produced water or that a substantial difference exists in the geological formation. Interestingly the remainder of the ratios are within an order of each other, especially the Ba/Sr, Ba/Na and Sr/Na ratios. This could possibly indicate that similar chemical properties in the produced water can be expected from the Karoo type shales in which the hydraulic fracturing events will take place. However, it should be kept in mind that without hydraulic fracturing field data these values can only be assumed to indicate possible chemical species. This clearly indicates that a test site should be established to determine the quantity and quality of the backflow water over an extended time period.

Table 5: Reported composition of shale samples obtained from various sources

Source	Element (mg/l)							
	Ba	Ca	Fe	Li	Sr	Mg	K	Na
Hem1	250	22500	38800	46	290	16400	24900	4850
Hayes2	1552	8451	64	70	1650	728	237	24043
Karoo3	2.7	2400	770	1	3.2	308	50	50

[i] - 1. Hem report USGS [35]; 2. Hayes report GTI [27]; 3. Karoo Sample leached in lab with HCl acid

Table 6: Ratios of chemical compositions from reported shale samples

Source	Element (mg/l)						
	Ba/Sr	Ba/Ca	Ba/Li	Ba/Mg	Ca/Mg	Ba/Na	Sr/Na
Hem1	0.86	0.01	5.43	0.02	1.37	0.05	0.06
Hayes2	0.94	0.18	22.17	2.13	11.61	0.06	0.07
Karoo3	0.84	0.01	2.70	0.01	7.79	0.05	0.06

[i] - 1. Hem report USGS [35]; 2. Hayes report GTI [27]; 3. Karoo Sample leached in lab with HCl acid

A recent sampling event took place at the Soekor core holes. Currently, the data set is limited and contains both the Soekor core hole data and surrounding well water. Interestingly, one of the core holes produced natural gas that could be ignited. The data is presented in Figure 7 in association with the Hayes report [27] data. Soekor data points are indicated as triangles, with SA 1, 5 and 7 representing samples from Soekor core holes. Sample data SA1 amd SA5 has a similar water type than that observed for the Hayes data set, which would indicate a highly mineralised water type. The main difference in the produced water is that the Soekor core holes have a reduced total dissolved solids content of approximately 6500-7200 mg/l. The third Soekor core hole water data (SA7) clearly has a Na/K-HCO_3 water type and a TDS of 440mg/l, indicating the presence of a surface aquifer interaction or a recharge mechanism that is introducing freshwater into the system. Furthermore, it is unsure at this stage if the anulus of the bore is still intact or if short-circuiting is taking place at the site. The data presented is only preliminary and further data sets is required to fully characterise these sites.

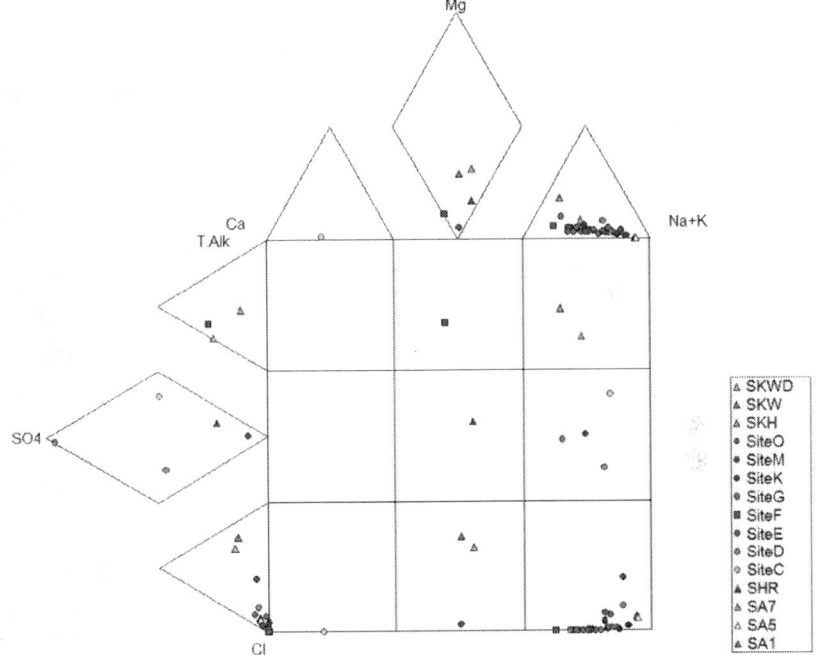

Figure 7: Expanded Durov diagram illustrating the different water types characterised from the Marcellus [27] and Soekor sites.

The Soekor core bores have been abandoned for nearly 40 years and there is still evidence that relatively high salinity water is produced from these sites. The rate of water production is relatively low compared to the data presented by Hayes [27], but as the production rate of water decreases at the sites it is currently unclear if there is still a hydraulic pressure that could produce water at surface. In the instance of the Soekor sites it does seem likely that recharge is occuring and that unless these holes are adequately sealed, a continuous discharge of water and gas might be possible.

Environmental Impacts of Hydraulic Fracturing

The concerns over hydraulic fracturing centre on a few main issues (Figure 8): (1) migration of gas, (2) migration of fracturing fluids, (3) water use, (4) management of produced water, (5) surface spills and

(6) identification of chemical additives. Each of these issues will be addressed in the following numbered sections, it is a summary of best practice guidelines to prevent uncontrolled releases of hydraulic fracturing fluid into the environment or to protect the environment within a reasonable limit of practice.

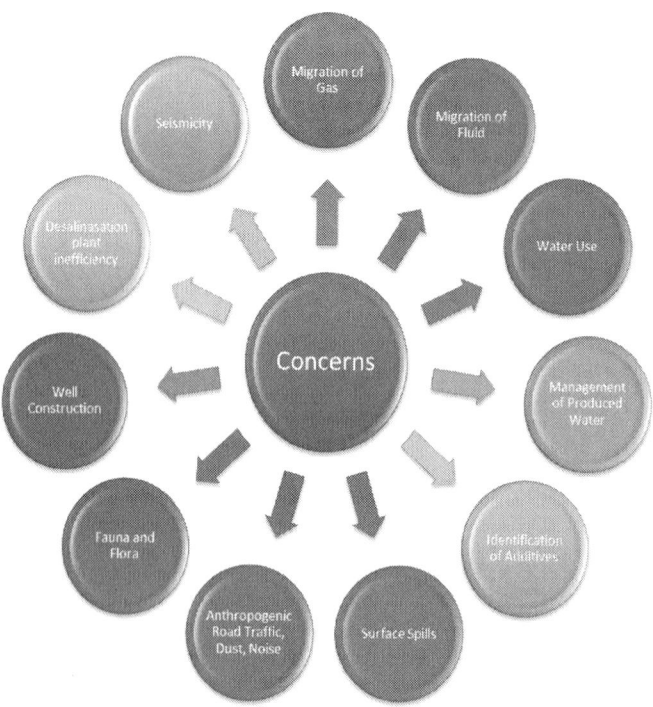

Figure 8: Main concerns regarding impacts of hydraulic fracturing on the environment.

Michigan's laws and rules effectively protect water and other natural resources as well as public health and safety from potential adverse effects of hydraulic fracturing. The Department of Environmental Quality (DEQ) has more than 50 staff employed in enforcing these state requirements. To date, only a few productive Utica/Collingwood Shale gas wells have been drilled in Michigan and the potential for more extensive development is unknown; however, the DEQ is taking a proactive approach in addressing large-scale hydraulic fracturing as well as other issues associated with deep shale gas development.

- Migration of Gas or Fracture Fluids. A major concern in natural gas development is the prevention of migration of gas or other fluids out of the reservoir and into overlying strata, particularly fresh water aquifers. In cases where this has occurred, it has been the result of well construction problems and not of hydraulic fracturing itself [36; 37]. At depths of about 610 meters or less, fractures propagate horizontally due to the natural stress regime of the rock. This confines the fractures to the gas reservoir. At greater depths, fractures may propagate vertically; however, characteristics of overlying rock layers prevent fractures from extending above the top of the gas reservoir. The installation of steel pipe ("casing"), encased in cement, is key to preventing migration of gas or fluids. Michigan regulations require that each oil and gas well have a casing and cementing plan that will effectively contain gas and other fluids within the wellbore, whether related to fracturing or not. Surface casing must be set a minimum of 35 meters into the bedrock and 35 meters below any fresh water zones and cemented from the base of the casing to the ground surface. Before fracturing or other operations can take place to complete a well for production, an additional string of production casing must be set to the depth of the reservoir and cemented in place. Depending on depth, additional protective casing may be required. To provide additional protection for aquifers and well integrity, the DEQ imposes a permit condition for wells in shallow reservoirs prohibiting hydraulic fracturing within 15 meters of the base of the surface casing. In addition, Instruction 1-2011 requires reporting of volumes, rates, and pressures (including pressure immediately outside of the pipe used to inject the fracturing fluid). Also, DEQ staff check wells in the vicinity to assure there are no wells or other features that could serve as conduits for unwanted movement of fracturing fluids.
- Water Use. A fracture treatment of a typical Antrim gas well requires about 189 m^3 of water. In the emerging Utica/Collingwood Shale gas development, the amount of water needed to fracture a horizontal well may be up to 18 927 m^3 or more. To put this in perspective, 18 927 m^3 is the volume of water typically used by eight to ten acres of corn during a growing season. Withdrawal of water for oil and gas operations

is exempt from the requirements of Michigan's water withdrawal statute; however, Instruction 1-2011 requires the operator to perform the same water withdrawal impact assessment as any other user of large volumes of water. It also requires installation and monitoring of an observation well if there is a freshwater supply well within one-quarter mile. The DEQ will not approve a withdrawal of water for hydraulic fracturing if it is likely to cause a significant adverse impact to groundwater or surface water.

- Management of Produced Water. Proper management of produced water is essential in protecting public health and the environment. In Michigan, produced water must be managed. Hydraulic Fracturing and disposed of according to strict rules specifically applying to those fluids. The fluids must be contained in steel tanks and transported to disposal wells where they are injected into deep rock layers that are isolated from fresh water supplies. The disposal wells are licensed by both the DEQ and the U.S. Environmental Protection Agency, and must be tested periodically to assure well integrity. Instruction 1-2011 requires reporting of the volume of flowback water recovered after a hydraulic fracturing operation.
- Surface Spills. Spills of chemical additives or flowback water can have adverse environmental or public health impacts. Michigan requires secondary containment under tanks, wellheads, and other areas where spills may be most likely. If a spill does occur, it must be reported immediately to the DEQ, and all spills must be promptly recovered and cleaned up according to strict requirements.
- Identification of Chemical Additives. Instruction 1-2011 requires oil and gas operators to provide to the DEQ copies of all Material Safety Data Sheets (MSDSs) for additives used in hydraulic fracturing. The MSDSs include information on physical characteristics, toxicity, health effects, first aid, reactivity, storage, disposal, protective equipment, and spill response. The DEQ will post the MSDSs on the Department's web site for public review. While the details on some of the chemical compounds used in hydraulic fracturing are exempted from disclosure on the MSDSs under federal law, the MSDSs will provide enough information for the DEQ to track and monitor spills.

The regulations enforced in Michigan was designed for the state specifically, in the instance of South Africa the following key differences will need to be considered.

- It most likely will not be possible to dispose of brine by re-injection into deep wells unless an exception in relation to the Water Act is obtained. This will introduce another issue which is disposal of solids and brines that is produced from water purification processes.
- The volume of material produced over the lifetime of a well field might require some engineering adaptation and/or disposal in dedicated waste storage facility constructed just for this purpose. It is still an open question as to how this will be managed.
- On the issue of water use, there is currently enough usable water available to proceed with hydraulic fracturing in the Karoo basin; but it will require planning and development of small scale well fields to abstract adequate volumes.
- Desalination plant efficiencies will need to be increased as the systems are currently sensitive to inflow water quality. It will most likely be associated with a multi-stage facility to remove organics (BTEX, PAHs) from the produced water and total dissolved salts. The composition of the salts is assumed to be mostly Na/Cl but it is expected that Ca, Fe and Mg salts will also be present. The presence of Fe salts might also pose interesting processing challenges for these plants.
- The presence of dolerite formations and thermal springs indicate that there might be a possible upward migration pathway for contamination migration. The probability of this occurring in the vicinity of the well field cannot be ruled out; especially if control measures and well field integrity is not measured over the lifetime of the well.
- In addition post-closure monitoring should be conducted to ensure that well failure does not cause upward migration of contaminants (i.e. Soekor sites).

CONCLUSIONS

South Africa has in the past been heavily dependent on its rich coal resources to supply it of electricity and fuel; with the discovery of an unconventional terrestrial gas resource it is currently entering a new age of energy independence. The development of this resource has put a strain on local communities due to fears of contaminated surface water and groundwater resources. The area currently being investigated, has both a historical and national significance and emotions are running high. Due to the sensitivity of South Africans regarding the Karoo, a great deal of care is required when gas exploration and eventual development occurs in this area. Key concerns is that the environment will be impacted to such an extent that it will be irrevocably changed. The geology of the area is to a certain extent complex and has dolerite sills and dykes which intrude the country rock. However, the Ecca formations of the Karoo has a considerable carbon content and suitable thickness to make it an ideal target for shale gas development. In this paper the process of hydraulic fracturing have been investigated from a hydrochemical perspective. Firstly, the composition of hydraulic fracturing fluids and the possible risks it pose to the surface and subsurface systems. Secondly, backflow water was evaluated for the Marcellus Shale since no current hydraulic fracturing program has been initiated in South Africa to target the Ecca shale formations. A summary of the key parameters were discussed as well as the production of flowback water and salt loading. Issues relating to salt loading were mainly related to treatment plants and the ability to effectively dispose of the produced brines and salts. A limited set of samples were incorporated into this paper from the Soekor core holes, and similar trends in water type was observed for both the Soekor sites and Marcellus samples.

Environmental impacts due to hydraulic fracturing activities were discussed. Due to South Africa's recent introduction to unconventional gas development a number of important regulatory processes does not exist, i.e. well and site inspectors. The state of Michigan's pro-active approach to regulating shale gas development addressed most of the issues which will be prevelent in the South African regulators mind. Finally, key differences between the regulatory environments were presented as well as unique challenges that faces South Africa in developing the unconventional gas resource.

ACKNOWLEDGEMENTS

We would like to acknowledge the University of the Free State and Water Research Commission of South Africa for funding. Dr. L. Chevallier for the geological information and assistance in strata characterisation.

REFERENCES

1. T Twine, M Jackson, R Potgieter, D Anderson, and L Soobyah, Karoos Shale Gas Report: Special Report on Economic Considerations Surrounding Potential Shale Gas Resources in the Southern Karoo of South Africa. In, (Econometrix), 76. Econometrix Park, 8 West Street, Houghton, Johannesburg, 2198; 2012
2. G Steyl, G. J Van Tonder, and L Chevallier, State of the Art: Fracking for Shale Gas Exploration in South-Africa and the Impact on Water Resources. In, (Commission W.R.), 96. Pretoria; 2012
3. M. R Xiphu, S. R Mills, L Chevallier, J Marot, M Mkhize, T Motloung, P Ngesi, A Okonkwo, M Msmart, M Solomons, A Tiplady, and E Van Wyk, Report on Investigation of Hydraulic Fracturing in the Karoo Basin of South Africa. In, (Resources D.o.M.), 96. 70 Meintjies Street, Sunnyside: Department of Mineral Resources; 2012
4. G. C Howard, and C. R Fast, Hydraulic Fracturing.. In Monograph of the Henry L. Doherty Series. New York: Society of Petroleum Engineers; 1970
5. C. T Montgomery, and M. B Smith, Hydraulic Fracturing-History of an Enduring Technology. Journal of Petroleum Technology 62: 16; 2010
6. H. K Van Poolen, J. M Tinsley, and C. D Saunders, Hydraulic Fracturing-Fracture Flow Capacity Vs. Well Productivity, Trans. AIME 21391951958
7. D. M Rowsell, and A. M. J Swardt, Diagenesis in Cape and Karoo Sediments, South Africa, and Its Bearing on Their Hydrocarbon Potential. Trans. Geol. Soc. S. Afr. 79811451976

8. B. P Tissot, and D. H Welte, Petroleum Formation and Occurrence. New York: Springer-Verlag; 1984
9. J Hayden, and D Pursell, The Barnett Shale, Visitors Guide to the Hottest Gas Play in the Us. In, (INC P.E.52Houston, Texas: Institutional Research; 2005
10. M. J De Wit, The Great Shale Debate in the Karoo. South African Journal of Science 107: 9; 2011
11. J. J Veevers, D. I Cole, and E. J Cowan, Southern Africa: Karoo Basin and Cape Fold Belt. In Permian-Triassic Pangean Basins and Foldbelts Along the Panthalassan Margin of Gondwanaland, (Veevers J.J. and Powell C.M.). Boulder, Colorado: Geological Society of America; 1994
12. F. M Gradstein, J. G Ogg, A. G Smith, W Bleeker, and L. J Lourens, A New Geologic Time Scale, with Special Reference to Precambrian and Neogene. Episodes 27831002004
13. J. H. A Viljoen, Tierberg Formation. In Catalogue of South African Lithostratigraphic Units, (Johnson M.R.). Pretoria: South African Committee for Stratigraphy, Government Printer; 2005
14. D. I Cole, and I. R Mclachlan, Oil Shale Potential and Depositional Environment of the Whitehill Formation in the Main Karoo Basin. Geological Survey of South Africa; 1994
15. D. I Cole, and A. D. M Christie, A Palaeoenvironmental Study of Black Mudrock in the Glacigenic Dwyka Group from the Boshof-Hertzogville Region, Northern Part of the Karoo Basin, South Africa. In Earth's Glacial Record. International Geological Correlation Project 260, (Deynoux M., Miller J.M.G., Domack E.W., Eyles N., Fairchild I.J. and Young G.M.). Cambridge: Cambridge University Press; 1994
16. D. I Cole, and W Basson, Whitehill Formation. In Catalogue of South African Lithostratigraphic Units, (Johnson M.R.). Pretoria: South African Committee for Stratigraphy, Government Printer; 1991
17. D. I Cole, Prince Albert Formation.. In Catalogue of South African Lithostratigraphic Units, (Johnson M.R.). Pretoria: South African Committee for Stratigraphy. Pretoria: Government Printer; 2005
18. M. R Johnson, C. J Van Vuuren, J. N. J Visser, D. I Cole, H. D. V Wickens, A. D. M Christie, D. L Roberts, and B. L.G. Sedimentary

Rocks of the Karoo Supergroup. In The Geology of South Africa, (Johnson M.R., Anhaeusser C.R. and Thomas R.J.). Johannesburg/Pretoria: Geological Society of South Africa/Council for Geoscience; 2006

19. T Branch, O Ritter, U Weckmann, R. F Sachenhofer, and F Schilling, The Whitehill Formation-a High Conductivity Marker Horizon in the Karoo Basin. South African Journal Geology 1104654762007

20. GA. Background Information Document and Invitation to Comment. In Proposed South Western Karoo Bsoin Gas Exploration Project, (B.V. S.E.C.). 2011.

21. C Dondo, L Chevallier, A. C Woodford, R Murray, and L Nhleko, Flow Conceptualisation, Recharge and Storativity Determination in Karoo Aquifers, with Special Emphasis on Mzimvubu-Keiskamma and Mvoti-Umzimkulu Water Management Areas in the Eastern Cape and Kwazulu-Natal Provinces of South Africa. In Research Reports. Pretoria: Water Research Commission of South Africa; 2010

22. D. M Rowsell, and J Connan, Oil Generation, Migration and Preservation in the Middle Ecca Sequence near Dannhauser and Wakkerstroom. In Some Sedimentary Basins and Associated Ore Deposits of South Africa. Special Publication of the Geological Society of South Africa, (Anderson A.M. and Van Biljon W.J.). Pretoria: Geological Society of South Africa; 1979

23. G Steyl, Estimation of Representative Transmissivities of Heterogeneous Aquifers. In Institute for Groundwater Studies, 132. Bloemfontein: Free State University; 2012

24. R Murray, K Baker, P Ravenscroft, C Musekiwa, and R Dennis, A Groundwater Planning Toolkit for the Main Karoo Basin. In Research Reports. Pretoria: Water Research Commission of South Africa; 2012

25. L. E Kent, The Thermal Waters of the Union of South Africa and South West Africa. Transactions of the Geological Society of South Africa 522312641949

26. J. H. A Viljoen, F. D. J Stapelberg, M Cloete, Technical Report on the Geological Storage of Carbon Dioxide in South Africa. In. Pretoria: Council for Geoscience South Africa; 2010

27. T Hayes, Sampling and Analysis of Water Streams Associated with the Development of Marcellus Shale Gas. In, 356. Des Plaines, IL 60018: Marcellus Shale Coalition; 2009
28. HalliburtonHydraulic Fracturing: Fluids Disclosure. In, Description of Hydraulic Fracturing and Fluids Used in the Process 28 February. 2013
29. FracFocusFind a Well. In, Find a Well (Hydraulic Fracturing)28 February. 2013
30. SWEPILPHydraulic Fracturing Fluid Product Component Information Disclosure. In. Erikson 448 4H; 2012
31. H. A Waxman, E. J Markey, and D Degette, Chemicals Used in Hydraulic Fracturing. In, (Representatives H.o.). Washington, USA: US Government; 2011
32. EIDA Fluid Situation, Typical Solution Used in Hydraulic Fracturing. In DOE, GWPC: Modern Gas Shale Development In the United States: A Primer (2009Energy Indepth; 2010.
33. L. A Fortson, B Yatzor, and T Bank, Physical and Chemical Associations of Uranium and Hydrocarbons in the Marcellus Shale. In Proceedings of' Northeastern (46th Annual) and North-Central (45th Annual) Joint Meeting. 60Geological Society of America.
34. DOOGMC Department Office of Oil and Gas Management Compliance Report. In. Pennsylvania: Pennsylvania Department of Environmental Protection; 2013
35. J. D Hem, Study and Interpretation of the Chemical Characteristics of Natural Water. In, Contract 2254Alexandria: United States Geological Survey; 1985
36. K Bybee, Cement Design for the Life of the Well. Journal of Petroleum Technology 54, 8860612002
37. H. R Roshan, S.S. A Fully Coupled-Poroelastic Analysis of Pore Pressure and Stress Distribution around a Wellbore in Water Active Rocks. Rock Mech Rock Eng 441992102011

Chapter 8

Microbial Hydrocarbon Degradation: Efforts to Understand Biodegradation in Petroleum Reservoirs

Isabel Natalia Sierra-Garcia[1]
and Valéria Maia de Oliveira[1]

[1]Microbial Resources Division, Research Center for Chemistry, Biology and Agriculture (CPQBA), University of Campinas, Campinas, Sao Paulo, Brazil

INTRODUCTION

The understanding of the phylogenetic diversity, metabolic capabilities, ecological roles, and community dynamics taking place in oil reservoir microbial communities is far from complete. The interest in studying microbial diversity and metabolism in petroleum reservoirs lies mainly but not only on providing a better comprehension of biodegradation of crude oils, since it represents a worldwide problem for petroleum industry. Generally, biodegradation of oil affects physical and chemical properties of the petroleum, resulting in a decrease of its hydrocarbon content and an increase in oil density, sulphur content, acidity and viscosity, leading to a negative economic consequence for oil production and refining operations [1,2]. Another important point for studying biodegradation lies on its important role in the global carbon cycle and the direct impact on bioremediation of polluted ecosystems. Furthermore, many of the enzymes involved in the degradation pathways are considered key catalysts in industrial biotechnology [3].

Despite these motivations and long recognition of petroleum as a the most important "primary energy" source, at present, microorganisms and factors involved in biodegradation of crude oil hydrocarbons in petroleum reservoirs are still not fully understood. The inaccessibility and complex microbiological sampling of petroleum reservoirs as well as the inherent limitations of the traditional culturing methods conventionally employed can explain this fact. Culture-based techniques have traditionally been the primary tools utilized for studying the microbiology of terrestrial and subsurface environments [4], which allowed the recovery and documentation of a large collection of bacteria capable of hydrocarbon utilization. Studies of numerous aerobic and anaerobic bacterial isolates have revealed mechanisms, which allow them to degrade specific classes of the highly diverse range of hydrocarbon compounds. Therefore, all we know about the degradation of petroleum compounds has come from studying isolated microorganisms. Here, we provide an overview of what is currently known about the mechanisms of aerobic and anaerobic degradation of hydrocarbons, as a result from biochemical and genomic approaches, we give a perspective of the petroleum microbial diversity unraveled so far, and finally we discuss the common oil reservoir characteristics that can be used to predict the most probable mechanism of degradation into deep petroleum reservoirs.

It is well known that microbial diversity in environment is several orders of magnitude higher than the one assumed based on previous cultivation methods [5]. A particularly large number of novel techniques have been developed, which now allow the determination of the *in situ* microbial diversity and activity on a particular site, screening for a particular gene or activity of interest, gene quantification, and DNA and mRNA sequencing and analysis from total communities. This book chapter will address how the implementation of such culture-independent molecular methods allow the access to the microbial diversity and metabolic potential of microorganisms and bring novel information about microbial diversity and new pathways involved in biodegradation processes taking place in petroleum reservoirs. This information will certainly contribute to a broader perspective of the biodegradation processes and corroborate with previous findings that degradation of pollutants in many cases is carried out by microbial consortia rather than a single species [6], where key species and catabolic genes are often not identical to those that have been isolated and described in the laboratory [7, 8].

MICROBIAL DIVERSITY IN OIL RESERVOIRS

Recognition of indigenous microbiota harbored by oil reservoirs has been discussed for a long time. Actually, determining the nature of isolated microorganisms from oil reservoirs (indigenous or nonindigenous) is a difficult issue concerning petroleum microbiologists. The reasons for this controversy rely mainly on the difficulty of aseptic sampling in deep oil reservoirs. This means that microorganisms observed in oil field fluids conceivably could be contaminants introduced during drilling operations and/or during sample retrieval, or could be material sloughed from biofilms growing in installed pipes. Another reason for skepticism is the commonplace practice of "water- flooding" (injection of surface waters or re-injection of natural formation waters to maintain reservoir pressure for oil production); since in this case microbes would be introduced during injection and therefore would not necessarily represent indigenous species [9].

In addition to this controversy, there is the fact that petroleum reservoirs are considered extreme environments where *in situ* conditions, like high pressure, temperature, salinity and anaerobic conditions, are considered as inhospitable to microbial activity. In fact, perception of deep subsurface as a sterile environment has only changed during the past two decades with the increasing awareness of the ability of microbes to colonize extreme environments. Actually, with the use of more sophisticated and appropriate sampling and cultivation techniques, as well as the application of molecular biological techniques to oil field fluids, the dogma of the sterile deep subsurface has been dispelled [9]. Rather, it has become clear that many oil reservoirs do harbor indigenous microbes (*e.g.* the genera *Geotoga* and*Petrotoga* isolated only from oil reservoirs) [10]. Nowadays it is clear that worldwide petroleum reserves are dominated by deposits that have been microbially degraded over geological time and biodegraded petroleum reservoirs represent the most dramatic manifestation of the deep biosphere [11].

In spite of the polemics on which micro-organisms would actually be native and which would be contaminants in oil reservoirs, a wide range of microbial taxonomic groups have been identified in oil reservoirs geographically distant using traditional techniques adapted to *in situ* conditions, as described by L'Haridon et al. [12], Grassia et al. [13] and reviewed by Magot et al [14], or combined with cultivation-independent molecular methods, as reported by Orphan et al. [15]. Table 1 summarizes the various physiological and taxonomical groups and species that have been isolated from oil reservoirs.

ASPECTS FROM OIL RESERVOIR DETERMINING MICROBIAL DEGRADATION

For a long time, the mechanism considered to be prevalent for oil degradation in petroleum reservoirs was the well documented aerobic microbial metabolism and it has long been thought that the flow of oxygen through meteoric waters was necessary for in-reservoir petroleum biodegradation [16]. This mechanism has been widely accepted despite the fact that oxygen would likely be consumed by oxidation of organic matter in near surface sediments and therefore, would be very unlikely for oxygen to reach deep petroleum reservoirs [11].

Recently, the discovery of the ability of microorganisms to degrade anaerobically hydrocarbon oil components and the detection of metabolites characteristic of anaerobic hydrocarbon degradation in oil samples from biodegraded reservoirs, but not in non-degraded reservoirs or aerobically degraded oils [11], have provided valuable information to determine the processes involved in the degradation of oil reservoirs. Nowadays, evidences of such degradation through anaerobic rather than aerobic processes are becoming more substantial and compelling [17].

It is known that microorganisms in anaerobic conditions can use a variety of final electron acceptors, including nitrate, iron, sulfate, manganese and, more recently, chlorate. Anaerobic degradation has also been coupled to methanogenesis, fermentation and phototrophic metabolism but growth of these microorganisms and, therefore, biodegradation rates are significantly lower compared to aerobic degraders. These anaerobic processes have been demonstrated in surface sediments and pure cultures or enrichments in laboratories [18] and all of them potentially play a role in oil biodegradation in anoxic petroleum reservoirs [11]. However, nitrate, like oxygen, is highly reactive and would likely be completely consumed before it could reach the oil reservoir [17]. In deep reservoirs, the supply of large amounts of Fe(III) or manganese(IV) via meteoric water influx are unlikely due to poor solubility and slow water recharge rates in subterranean cycles. Therefore, iron and manganese, which could be used as electro acceptors for oil oxidation, are unlikely to be responsible for significant compositional changes in the oil, considering their limited availability in the reservoir. Accordingly, oil degradation linked to sulfate reduction and methanogenic would therefore explain the consistent hydrocarbon compositional patterns seen in degraded oils worldwide [17]. Sulfate arises from geological sources, such as evaporitic sediments and limestone, or from the injection of seawater for pressure stabilization, and may lead to significant oil degradation and increased residual-oil sulfur content. Methanogenic oil degradation, on the other hand, does not require external electron acceptors and leads to less overall souring of the oil reservoir. Several studies have described *in vitro*methanogenic degradation of crude oil related compounds [19, 20] Jones et al., 2008), including n-alkanes [21, 20] and aromatic hydrocarbons [17].

Table 1: Summary of bacteria isolated from oil reservoirs worldwide

Organism	Taxonomical group	Metabolism	Origin	Reference
Thermodesulforhabdus norvegicus	Deltaproteobacteria	Sulfate-reducer	Oil field in Norway	[22]
Desulfacinum infernum	Deltaproteobacteria	Sulfate-reducer	North see petroleum reservoir near Scotland	[23]
Desulfomicrobium norvegicum	Deltaproteobacteria	Sulfate reducer	Petroleum reservoir in Canada	[24]
Desulfovibrio sp.	Deltaproteobacteria	Sulfate reducer	Petroleum reservoir in Canada	[24]
Dethiosulfovibrio peptidovorans	Bacteria, Synergistetes	Sulfate reducer	Oil well in the Emeraude oilfield in Congo, Central Africa,	[25]
Desulfotomaculum thermocisternum	Bacteria, Firmicutes	Sulfate reducer	Oil reservoir in the North sea	[26]
Deferribacter sp.	Bacteria, Deferribacteres	Sulfate reducer	California oil fields	[15]
Halanaerobium congolense	Bacteria, Firmicutes	Thiosulfate- and sulfur-reducing bacterium	African oil field	[27]
Thauera phenylacetica	Betaproteobacteria	Nitrate reducer	Petroleum reservoir in Canada	[24]
Pseudomonas stutzeri	Gammaproteobacteria	Nitrate reducer	Petroleum reservoir in Canada	[24]
Garciella nitratireducens	Bacteria, Firmicutes	Nitrate reducer	Oil field in Tabasco, Gulf of Mexico	[28]
Geobacillus subterraneus, Geobacillus uzenensis	Bacteria, Firmicutes	Nitrate reducer	Petroleum reservoir in China	[29]
Lactosphaera pasteurii	Bacteria, Firmicutes	Fermentative	Petroleum reservoir in Canada	[24]

Propionicimonas paludicola	Bacteria, Firmicutes	Fermentative	Petroleum reservoir in Canada	[24]
Anaerobaculum	Bacteria, Synergistetes	Fermentative	California oil fields	[15]
Thermococcus sp.	Archaea, Euryarchaeota	Fermentative	California oil fields	[15]
Thermococcus sibericus	Archaea, Euryarchaeota	Fermentative	Petroleum reservoir in Western Siberia	[30]
Petrotoga sp.	Bacteria, Thermotogae	Fermentative	California oil fields	[15]
Petrotoga olearia; P. siberica	Bacteria, Thermotogae	Fermentative	Petroleum reservoir in Western Siberia	[12]
Thermoanaerobacter	Bacteria, Firmicutes	Fermentative	California oil fields	[15]
Thermotoga sp.	Bacteria, Thermotogae	Fermentative	California oil fields	[15]
Thermosipho geolei	Bacteria, Thermotogae	Fermentative	Petroleum reservoir in Western Siberia	[12]
Anaerobaculum thermoterrenum	Bacteria, Synergistetes	Fermentative	Oil well in Utah	[23]
Fusibacter paucivorans	Bacteria, Firmicutes	Fermentative	Oil well in the Emeraude oilfield in Congo, Central Africa	[31]
Thermovirga lienii	Bacteria, Synergistetes	Fermentative	Oil reservoir in the North sea	[32]
Methanococcus	Archaea, Euryarchaeota	Methanogen	California oil fields	[15]
Methanococcus thermolithotrophicus	Archaea, Euryarchaeota	Methanogen	North sea old field in Norway	[33]
Methanoculleus	Archaea, Euryarchaeota	Methanogen	California oil fields	[15]
Methanobacterium	Archaea, Euryarchaeota	Methanogen	California oil fields	[15]

Deep subsurface environments such as petroleum reservoirs are logistically much more difficult to study than contaminated shallow subsurface environments [17]. Since in many biodegraded petroleum reservoirs most biodegradation occurs close to the oil water transition zone, it has been proposed that the oil–water transition zone (OWTZ) provides suitable physical and chemical conditions for microbial activity [17].

There are other physical and chemical parameters influencing *in situ* biodegradation. Temperature is one of the main factors which limits oil degradation in reservoir, and, empirically, it has been repeatedly observed that biodegradation does not occur in oil reservoirs with *in situ* temperatures >80-90°C [34]. Salinity is another factor that affects in-reservoir oil biodegradation, especially in combination with temperature [13]. Typically, reservoirs with highly saline waters show limited oil biodegradation [11]. This is consistent with the observations that it has not been possible to cultivate microorganisms from reservoir waters with salinity greater than 100 g/L [13]. Pressure seems to be a less limiting factor, except that it may select for certain physiological types and influences the pH of pore waters by increasing dissolution of CO_2 [9]. The availability of electron donors and acceptors governs the type of bacterial metabolic activities within oil field environments [14]. The potential electron donors include CO_2, hydrocarbons, H_2 and numerous organic molecules. Availability of fixed nitrogen is unlikely to limit microbial activity in reservoirs. However, the availability of water-soluble nutrients, like phosphorus and/ or oxidants (terminal electron acceptors such as ferrous iron, sulfate or CO2), is more likely to limit *in situ* microbial activity [9]. Nonetheless, physiological characteristics of microorganisms indigenous to petroleum reservoirs shed light on the conditions under which petroleum degradation may occur and the potential degradation mechanisms.

HYDROCARBON DEGRADATION

Hydrocarbons are understood as the compounds that consist exclusively of carbon and hydrogen. Because of the lack of functional groups, hydrocarbons are largely apolar and exhibit low chemical reactivity at room temperature. Differences in their reactivities are primarily determined by the occurrence, type and arrangement of unsaturated bonds. Therefore, in this chapter, we will use the common way to classify hydrocarbons according to their bonding features: i) aliphatic group, which includes straight-chain (n-alkanes), branched-chain and cyclic compounds and ii) aromatic group which includes mono or polycyclic hydrocarbons an many important compounds which also contain aliphatic hydrocarbon chains (e. g., alkylbenzenes).

Already a century ago, bacterial isolates had been reported to use aliphatic and aromatic hydrocarbons as sole carbon and energy sources [35]. Since then, numerous aerobic, and also anaerobic, bacterial isolates have been studied in order to understand the mechanisms which allow them to degrade specific members of the highly diverse aliphatic and aromatic compounds. Degradation by such isolates has been investigated thoroughly and results have revealed that they can completely degrade most classes of hydrocarbons, including alkanes, alkenes, alkynes and aromatic compounds. Such degradation can occur aerobically, with oxygen, or anaerobically, with nitrate, ferric iron, sulfate or other electron acceptors [36].

Efforts to overview the metabolism of hydrocarbons in microorganisms are confronted with the chemical diversity of such compounds and their reactivities, as well as with various microbial life styles [36]. The study of biodegradation is conventionally treated in separate areas: aliphatic vs. aromatic hydrocarbons, aerobic vs. anaerobic degradation pathways, physiology and overall metabolic pathways vs. enzymatic mechanisms and structures, often with limited knowledge and data exchange. Nonetheless, each of these study areas deals with the same central point that is the "metabolic challenge" to guide an apolar, unreactive compound composed only of carbon and hydrogen into the metabolism [36]. The hydrocarbon must be first functionalized and currently it has been recognized that there is a surprisingly diversity of reactions of activation that had evolved in microorganisms (Table 2).

Table 2: Overview of aerobic and anaerobic mechanisms for hydrocarbon activation in bacteria

Mechanisms for hydrocarbon activation		
	Aerobic	**Anaerobic**
Short-Chain non-methane alkanes C2-C10	• Non-heme iron monooxygenase similar to sMMO (C2-C9) • Copper-containing monooxygenase similar to pMMO (C2-C9) • Heme-iron monooxygenases (also refered as soluble Cytochrome P450 (C5-C12)	• Fumarate addition
Long-Chain alkanes >C10	• Heme-Monooxygenase (P450 type) • [Fe2]-Monooxygenase • Non-heme iron monooxygenase (AlkB-related) (C3-C13 or C10-C20) • Flavin-binding monooxygenase (AlmA) (C20- C36) • Thermophilic flavin-dependent monooxygenase (LadA) (C10-C30)	• Fumarate addition• Carboxylation
Aromatic hydrocarbons	• [Fe]-Dioxygenase • [Fe2]-Monooxygenase • [Flavin]-Monooxygenase	• Fumarate addition • Hydroxylation • Carboxylation

Mechanisms for hydrocarbon activation are basically different in aerobic and anaerobic microorganisms. Under oxic conditions, hydrocarbon metabolism is always initiated using molecular oxygen as a co-substrate in mono- or dioxygenase reactions that enable the terminal or sub-terminal hydroxylation of aliphatic alkane chains or the mono or dihydroxylation of aromatic rings [37]. In the hydrocarbon activation under anoxic conditions, some proposed reactions comprise: (1) addition to fumarate by glycyl-radical enzymes, (2) methylation of unsubstituted aromatics, (3) hydroxylation with water by molybdenum cofactor containing enzymes of an alkyl substituent via dehydrogenase, and (4) carboxylation catalyzed by yet- uncharacterized enzymes which may actually represent a combination of reaction (2) followed by reaction (1) [38; 37]. Although all these mechanisms of hydrocarbon anaerobic activation have been proposed, the required signature metabolites and enzymes involved have been characterized only for (1) addition to fumarate (demonstrated for toluene, xylene, ethylbenzene, methylnaphthalene, alkanes and alicyclic alkanes); for (3) hydroxylation (demonstrated for ethylbenzene); and for (4) carboxylation (demonstrated for benzene and naphtalene) [39].

BIOCHEMICAL AND GENETIC PATHWAYS OF MICROBIAL HYDROCARBON DEGRADATION

The enzymatic reactions involved in the aerobic degradation of hydrocarbons by bacteria have been extensively studied for several decades [37]. Genes encoding enzymes for degradation are relatively well understood for aerobic and easily cultivable microorganisms, particularly for a *Pseudomonas* strain, known as *P. putida* GPo1, as well as for the strains *Acinetobacter* sp. ADP1 and *Mycobacterium tuberculosis* H37Rv [39, 40]. On the other hand, the anaerobic hydrocarbon degradation has gained more attention since is supposed to be the predominant mechanism occurring in several polluted environments and oil reservoirs. However, its study is an incipient area because of the peculiarities of the reservoir environment and difficulties that arise from attempts to characterize these communities. Nevertheless, several bacteria from other environments able to use alkanes as carbon source in the absence of oxygen have been described in the last few years [41], but anaerobic bacteria able to degrade hydrocarbons under conditions found in deep petroleum reservoirs have not been isolated so far [2]. Figure 1 represents an overview of the main mechanisms and pathways used by microorganisms to degrade hydrocarbon compounds under aerobic and anaerobic conditions.

Aerobic Degradation

Aliphatic Hydrocarbons

In most degradation pathways described, the substrate n-alkane is oxidized to the corresponding alcohol by substrate-specific terminal monooxygenases/hydroxylases. The alcohol is then oxidized to the corresponding aldehyde, and finally converted into a fatty acid. Fatty acids are conjugated to CoA and subsequently processed by β – oxidation to generate acetyl-CoA [42, 40]. Subterminal oxidation has also been described for both short and long-chain alkanes [40]. Both terminal and sub-terminal oxidation can coexist in some microorganisms [41]. Initial terminal hydroxylation of n-alkanes in bacteria can be carried out by enzymes belonging to different classes, named: (1) propane monooxygenase (C3), (2) different classes of butane monooxygenase (C2-C9), (3) CYP153 monooxygenases (C5-C12), (4) AlkB-related non-heme iron monooxigenase (C3-C10 or C10-C20), (5) flavin-binding monooxigenase AlmA (C20-C36), (6) flavin-dependent monooxygenase LadA (C10-C30), (7) copper flavin-dependent dioxygenase (C10-C30) [43].

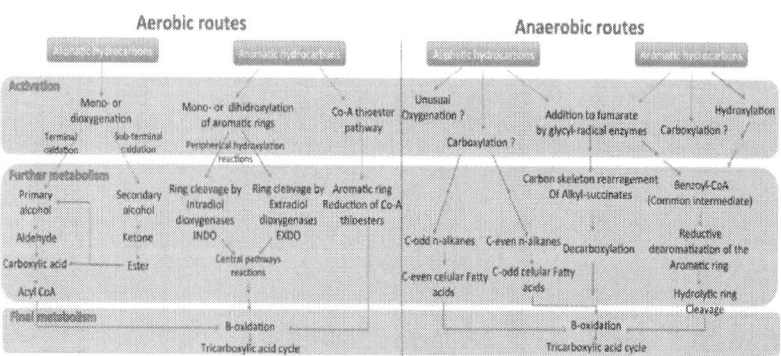

Figure 1: Pathways for aerobic and anaerobic bacterial degradation of hydrocarbon compounds. Two arrows represent more than one reaction.

Among all the alkane activating enzymes, the integral membrane non-heme iron monooxygenase (AlkB) is the best characterized one. Microorganisms degrading medium (C5-C11) and long (>C12)-length alkanes have been frequently related to the presence of *alk*B genes and that is why the presence of such genes have been widely used as functional biomarker for the characterization of aerobic alkane-degrading bacterial populations in several environmental samples [44, 45] and in bioremediation experiments [46, 47]. The degradation pathway of the *alk* system was first described in *Pseudomonas putida* GPo1 (formerly identified as *P. oleovorans* GPo1), where it is located on the OCT plasmid. In this model system, OCT plasmid contains two operons: *alk*BFGHJKL and *alk*ST [48]. The first operon encodes two components of the *alk* system, a particulate non-heme integral membrane alkane monooxygenase (AlkB) and the soluble protein rubredoxin (AlkG), as well as other enzymes involved in further steps. The second operon encodes for a rubredoxin reductase (AlkT and AlkS), which regulates the expression of the *alk*BFGHJKL operon [48, 49]. Since this system was described, AlkB homologous have been found in many alkane-degrading α- β - and γ -Proteobacteria and high G + C content Gram-positive bacteria (Actinobacteria) [39] and an increasing collection of alkane hydroxylase gene sequences has allowed the diversity analysis of hydrocarbon-degrading microbial populations in different ecosystems. However, comparisons of cloned *alk*B genes or gene fragments have showed that sequence diversity is very high, even among *alk*B genes within the same species [50].

In despite of the relevance of *alk*B genes as a functional biomarker of alkane-degrading bacterial communities, knowledge on the presence and diversity of *alk*B genes in oil reservoirs is scarce. Tourova et al. [51] analysed *alk*B diversity in thermophilic bacterial strains of the genus *Geobacillus* isolated from oil reservoirs or hot springs. They detected, for the first time, sets of *alk*B gene homologous in thermophilic bacteria, and some strains showed different homologous within the same genome. This fact was explained by the occurrence of horizontal gene transfer among these bacteria. Recently, Li et al. [52] aimed to evaluate *alk*B gene diversity and distribution in production water from 3 oilfields in China through a specific PCR-DGGE method. Results showed that sequences found in the water samples were similar to *alk*B genes from other corresponding alkane-degrading strains. But at the same time, they showed the presence of a considerable genetic diversity of alkB genes in the wastewater as evidenced by a total of 13 unique DNA bands detected. Studies on the degradation of alkanes in oil reservoirs are currently in a start point, but in the future they certainly will help to understand the process of degradation in oil reservoir.

In comparison to the few efforts in studying *alk*B system in oil reservoirs, much less is known about the presence of the other enzymatic systems previously listed, which have been described for aerobic degradation of n-alkanes in isolated bacteria or laboratory microcosms. For the most recent elucidated systems for alkane oxidation, named *almA* and *ladA* genes, nothing is known about the environmental distribution of these type of genes in petroleum contaminated sites [53] or oil fields, although the LadA complete degradation pathway has been characterized through genome and proteome analysis of *Geobacillus thermodenitrificans* NG80-2, a thermophilic strain isolated from a deep oil reservoir in Northern China [54]. Currently, it is believed that there are enzyme systems for alkane degradation which have still not been characterized and that may include new proteins unrelated to those already known [41]. Moreover, in many alkane degraders more than one alkane oxidation system have been observed, which have been reported exhibiting overlapping substrate ranges [39, 40]. These observations point out that in order to characterize and explore metabolic diversity and functions involved in alkane degradation one should take into consideration the high diversity of enzymes capable of initiating such metabolism.

Aromatic Hydrocarbons

The aerobic bacterial catabolism of aromatic compounds involves a wide variety of peripheral pathways that activate structurally diverse substrates into a limited number of common intermediates that are further cleaved and processed by a few central pathways to the central metabolism of the cell [55]. Metabolic pathways and encoding genes responsible for the degradation of specific members of a highly diverse range of aromatic compounds have been characterized for many isolated bacterial strains, predominantly from the Proteobacteria and Actinobacteria phyla [56]. Degradation by such isolates is typically initiated by members of one of the three superfamilies: the Rieske non-heme iron oxygenases (RNHO), the flavoprotein monooxygenases (FPM) and the soluble diiron multicomponent monooxygenases (SDM). Further metabolism is achieved through di- or trihydroxylated aromatic intermediates. Alternatively, activation is mediated by CoA ligases where the formed CoA derivates are subjected to selective hydroxylation [58, 53]. In the case of hydrophobic pollutants, such as benzene, toluene, naphthalene, biphenyl or polycyclic aromatics, aerobic degradation is usually initiated by activation of the aromatic ring through oxygenation reactions catalyzed by RNHO enzymes or, as intensively described for toluene degradation, through members of SDM enzymes [56].

Further intermediates can be catalyzed by two kinds of enzyme, intradiol and extradiol dioxygenases, which represent two classes of phylogenetically unrelated proteins [58]. These enzymes are key enzymes in the degradation of aromatic compounds, and many of such proteins and their encoding sequences have been described, purified and characterized in the last decades [56]. While all intradiol dioxygenases described so far belong to the same superfamily, the extradiol dioxygenases include at least three members of different families. Type I extradiol dioxygenases (e.g. catechol 2,3-dioxygenases and 1,2-dioxygenases) belong to the vicinal oxygen chelate superfamily enzymes. Type II extradiol dioxygenases are related to LigB superfamily (e.g. protocatechuate 4,5-dioxygenases) and the type III enzymes belongs to the cupin superfamily (e.g. gentisate dioxygenases) [53]. However, members of novel superfamilies performing crucial steps in aromatic metabolic pathways are still being discovered [56, 53].

The knowledge of metabolic properties of isolates has allowed the monitoring of the ability of microorganisms to mineralize aromatic hydrocarbons in soils. Typically, these studies have used primers designed based on conserved gene regions and focused on RNHO or SDM as targets for initiating degradation, or on Extradiol dioxygenases (EXDO) cleaving the aromatic ring [59]. These studies range from those searching for a narrow range of genes similar or identical to those observed in type strains using non-degenerated primers to those searching for subfamilies of homologous genes using degenerated primers [59]. However, due to the immense heterogeneity of such enzymes [57], there will never be a pair of primers that will reliably cover the huge diversity of a catabolic gene family in nature [53].

Anaerobic Degradation

Aromatic Hydrocarbons

We have already described the main mechanism for degradation of aromatic compounds in aerobic conditions, where oxygen is not only the final electron acceptor but also co-substrate of two key processes: hydroxylation and cleavage of the aromatic ring by oxygenases. In contrast, in the absence of oxygen, microorganisms use a complete different pathway, based in reductive reactions to attack the aromatic ring [61].

The biochemistry of some anaerobic degradation pathways of aromatic compounds has been studied to some extent; however, the genetic determinants of all these processes and the mechanisms involved in their regulation are much less studied [55]. Recent advances in genome sequencing have led to the complete genetic information for six bacterial strains that are able to anaerobically degrade aromatic compounds using different electron acceptors and that belong to different taxonomic groups of bacteria: denitrifying betaproteobacteria, *Thauera aromatica* and *Azoarcus* sp. EbN1, two alphaproteobacteria, the phototroph *Rhodopseudomonas palustris* strain CGA009 and the denitrifying *Magnetospirillum magneticum* strain AMB-1, and two obligate anaerobic deltaproteobacteria, the iron reducer *Geobacillus metallireducens* GS-15 and the fermenter *Syntrophus aciditrophicus* strain SB [55]. It is worth remembering that, in recent years, important inferences and generalizations have been made about the genetics involved in hydrocarbon metabolism based on these isolated bacteria under conventional laboratory conditions. However, potential novel genes, enzymes and metabolic pathways responsible for degradation processes are probably harbored by yet uncultivated bacteria.

The best understood and apparently the most widespread of these anaerobic mechanisms is the radical-catalyzed addition of fumarate to hydrocarbons, yielding substituted succinate derivatives. This reaction has been recognized for the activation of several alkyl-substituted benzenes as well for n-alkanes [62]. However, understanding of this fumarate-dependent hydrocarbon activation is most advanced in the case of toluene. The key enzyme in this process is the enzyme benzylsuccinate synthase. All enzymes required for β-oxidation of benzylsuccinate are encoded by the *bbs* operon. Subsequent degradation of benzoyl-CoA proceeds via reductive dearomatization, hydrolytic ring cleavage, β-oxidation to acetyl-CoA units and terminal oxidation to Co_2 [63]. In contrast to the anaerobic metabolism of toluene, degradation of ethylbenzene (and probably other alkylbenzenes with carbon chain of at least 2) is entirely different, despite the chemical and structural similarities between the two compounds, and involves a direct oxidation of the methylene carbon via (S)-1-phenylethanol to acetophenone [55]. Ethylbenzene is anaerobically hydroxylated and dehydrogenated to acetophone, which is then carboxyled and converted to benzoylCoA as the first common intermediate of the two pathways [62].

Genetics of the enzymatic system have been only characterized for these two mechanisms for anaerobic hydrocarbon activation. Genes encoding pathways that involve fumarate addition are typically organized in two operons. One operon includes the three structural genes of the protein catalyzing fumarate addition and the other includes genes required for converting succinate derivates to benzoyl-CoA [64]. Gene sequences and organization are relatively conserved among nitrate-reducing bacteria but differ somewhat from those of the iron reducer *G. metallireducens* [64] and substantially from those of the hexane-degrading nitrate reducer strain HxN1 [65]. Hydrocarbon dehydrogenation pathway is also organized in two operons. One operon contains the structural genes for the first two reactions (ethylbenzene dehydrogenase and 1-phenylethanol dehydrogenase) and the other contains the structural genes for acetophone carboxylase [64].

Kane et al. [66] developed the first real-time polymerase chain reaction (PCR) method to quantify hydrocarbon utilizers based on *bss*A genes of nitrate-reducing Betaproteobacteria. Since then, there have been several additional studies investigating the presence and/or distribution of anaerobic hydrocarbon utilizers in anaerobic environments via functional gene surveys of *bss*A, extending the range of detectable hydrocarbon-degrading microbes to iron and sulfate-reducing Deltaproteobacteria and revealing partially novel, site specific degrader populations [67, 68]. Other *bss*A-based detection studies in impacted environments, as well as studies that combine field metabolomics and molecular tools, are described by other authors [69, 70, 71]. Despite of the role of benzylsuccinate synthase in aromatic hydrocarbon degradation and its use as a biomarker are well documented, there is no study on the presence of this gene in oil reservoirs.

Aliphatic Hydrocarbons

Anaerobic degradation of alkanes has not been extensively studied as for some aromatic compounds. The presumable reasons include the greater attention given to BTEX compounds (benzene, toluene, ethylbenzene and xylenes) because of their classification as priority pollutants [71], also the fact that anaerobic growth with n-alkanes is even slower than that with the alkylbenzenes, and finally the fact that long chain alkanes are poorly soluble and often prevents the cultivation of cells homogeneously in the medium [72]. However, anaerobic degradation of alkanes is equally relevant, since alkanes are quantitatively the most important hydrocarbon components of petroleum, and some are acutely toxic and difficult to remediate [71]. Several anaerobic bacteria capable of degrading n-alkanes with 6 or more carbons in length, particularly hexadecane (C16), using sulfate or nitrate as electron acceptors have been isolated [72, 73].

The two main mechanisms of anaerobic degradation of n-alkanes described involve unprecedented biochemical reactions that differ completely from those employed in aerobic hydrocarbon metabolism [73]. The first involves activation at the subterminal carbon of the alkane by the addition of fumarate, analogously to the formation of benzyl succinate during anaerobic degradation of toluene, however further reactions are completely different involving dehydrogenation and hydration [72]. Studies conducted with established axenic cultures have indicated that anaerobic metabolism of oil allkanes predominantly proceeds via addition of fumarate to the double bound [72]. Although alkylsuccinate metabolites have rarely been detected in oil reservoir fluids [74, 75], they have been reported in oil-contaminated environments as well as in oilfield facilities, where their detection is indicative of *in situ* microbial degradation of oil alkanes [71, 75]. Alkylsuccinic acids as intermediates of anaerobic alkane oxidation were first studied by Gieg and Suflita [76] when surveying these metabolites in aquifers contaminated with condensate gas, natural gas liquids, gasoline, diesel, alkanes and BTEX. They found alkylsuccinates originating from C3 to C11 alkanes, as well as putative metabolites originating from compounds with one degree of unsaturation, such as alkenes or alicyclic alkanes. Since this report, other studies have detected alkylsuccinate derivates in petroleum contaminated groundwater systems [76], coal beds [70] and oil fields [74, 77]. The formation of alkylsuccinates is catalyzed by a strictly anaerobic glycyl radical enzyme which has been termed as alkylsuccinate synthase or (1-methyl-alkyl)succinate synthase (Ass or Mas). The genes encoding Ass have recently been identified in the alkane degrading sulfidogenic bacteria *D. alkenivoras* AK-01 [78] and *Desulfoglaeba alkanedexens*ALDC^T [71], as well as in nitrate reducing strains HxN1 [65] and OcN1 [79], all affiliated to the Proteobacteria phylum [80]. Recently, Callaghan et al. [71] detected *ass*A genes in a propane-utilizing mixed culture and in a paraffin-degrading enrichment culture maintained under sulfate-reducing conditions. Despite of no genes for benzyl-and alkylsuccinate synthase were found when environmental metagenome datasets of uncontaminated sites were analyzed in Callaghan et al [71], the authors consider that *ass*A gene could be a useful biomarker for anaerobic alkane metabolism.

The second mechanism for alkane anaerobic degradation is the carboxylation, mainly developed from the growth pattern of the sulfate-reducing strain Hxd3 [81], tentatively named as *Desulfococcus oleovorans*. This strain differs from other alkane degraders for converting C-even alkanes into C-odd cellular fatty acids whereas growth on C-odd alkanes resulted in C-even cellular fatty acids [81, 72]. More recently, Callaghan et al. [82] suggested that a carboxylation-like mechanism analogous to the activation strategy previously proposed by So et al. [81] was the probable route for the anaerobic biodegradation of hexadecane in an alkane-degrading, nitrate-reducing consortium. However, in both cases, the hypothetical fatty acid intermediate (2-ethylalkanoate) that should result from the incorporation of inorganic carbon at C-3 of the alkane has never been detected. There is an on-going debate about this initial activation mechanism. From an energetic point of view, the carboxylation of alkanes is not feasible under physiological conditions, unless the concentration of the fatty acid (2-ethylalkanoate) is in the micromolar order of magnitude or less [80].

Other alternative activation mechanisms are proposed for the anaerobic degradation of alkanes. For instance, the mechanism referred as "unusual oxygenation" is used by the strain *Pseudomonas chloritidismutans* AW-1^T, that is assumed to produce its own oxygen via chlorate respiration used for subsequent metabolism of alkanes [60]. Other alternative mechanism considers that activation in the anaerobic methanogenic system may be initiated by an anaerobic hydroxylation reaction [83].

MECHANISMS INVOLVED IN OIL BIODEGRADATION IN PETROLEUM RESERVOIRS

From those microorganisms studied in oilfields, methanogens have received particular attention since they have been isolated and molecularly detected in both low- and high-temperature reservoirs [88,89]. Their physiological characteristics and potential activity possibly involved in methanogenesis occurring in oil reservoirs have been demonstrated [90]. Furthermore, recently, Jones et al. [20] provided evidence that the patterns of hydrocarbon degradation observed in biodegraded petroleum reservoirs were the result of methanogenic processes. Therefore, microbiological and biogeochemical investigations have indicated that methanogenesis is a widely distributed process in petroleum reservoirs, although still poorly understood [90]. Methanogenesis is the terminal process of biomass degradation. Acetate and hydrogen are the most important immediate precursors for methanogenesis, and are converted into methane by acetoclastic and hydrogenotrophic methanogens, respectively [91]. Acetate can also be a precursor for methanogenesis through syntrophic acetate oxidation coupled to hydrogenotrophic methanogenesis, which is mediated by syntrophic bacteria and methanogenic archaea [92, 93, 94, 95]. Interestingly, acetate is generally abundant in many petroleum reservoirs, at concentrations ranging between 0.3 and 20 mM [96] hence, acetate metabolism is considered an important methane production process in those environments [90].

Cultivation-dependent and -independent approaches have shown the presence of acetoclastic and hydrogenotrophic methanogens and putative syntrophic acetate-oxidizing bacteria in reservoirs [88, 89,102], indicating that there should be two different pathways of acetate metabolism in the environment, namely acetoclastic methanogenesis and syntrophic acetate oxidation coupled with hydrogenotrophic methanogenesis. Some previous studies suggested that syntrophic acetate oxidation was most likely to occur in petroleum reservoirs, based on molecular biological analysis [89] and thermodynamic calculations [98]. In Jones et al. [20], the composition of oil in microcosms exhibiting methanogenic oil degradation is compared to patterns observed in biodegraded oils from the Gullfaks field in the North Sea. Analysis of the methanogenic communities from oil-degrading microcosms revealed a strong selection for CO_2-reducing methanogens against acetoclastic methanogens, and gas isotope modeling also revealed that to match the d13C of methane and carbon dioxide from biodegraded petroleum reservoirs 75–92% of methanogenesis should be via the CO_2 reduction pathway [20, 11].

The reason why syntrophic acetate oxidation predominates over acetoclastic methanogenesis in oil reservoirs remains unclear. There is evidence from studies of oil contaminated aquifers that crude oil can have a detrimental effect on acetoclastic methanogenesis and, in situations where acetoclastic methanogenesis is inhibited, methanogenic alkane degradation via syntrophic acetate oxidation may be thermodynamically the most favorable alternative pathway [11]. Nonetheless, one recent report suggests that acetoclastic methanogenesis may predominate in some methanogenic oil-degrading systems [19]. Although there is currently great interest in how much each of the two pathways contributes to methane production in petroleum reservoirs, no studies are being conducted to address this question [90].

METAGENOMICS AS A TOOL FOR A BETTER COMPREHENSION OF BIODEGRADATION

As stated previously, cultivation-based methods have traditionally been utilized for studying the microbiology in oil fields and have yielded valuable information about microbial interactions and their relations with hydrocarbons [42]. However, nowadays, it is known that only a small fraction of the microbial diversity in nature (1-10%) can be grown in the laboratory [84, 85, 86]. Therefore, it is assumed that the ecological functions of the majority of microorganisms in nature and their potential applications in biotechnology remain obscure [87].

In metagenomics, total DNA is extracted from appropriately chosen environmental samples, propagated in the laboratory by cloning techniques, submitted to sequence or function-based screenings and/ or subjected to large-scale sequence analysis (Fig. 2). Functional screening of metagenomic libraries offer the advantage that it does not rely on sequence homology to known genes, and for this reason, has allowed the isolation of different enzyme classes from several environments. The probability (hit rate) of identifying a certain gene depends on multiple factors that are intrinsically linked to each other: the host–vector system, size of the target gene, its abundance in the source metagenome, the assay method, and the efficiency of heterologous gene expression in a surrogate host [99].

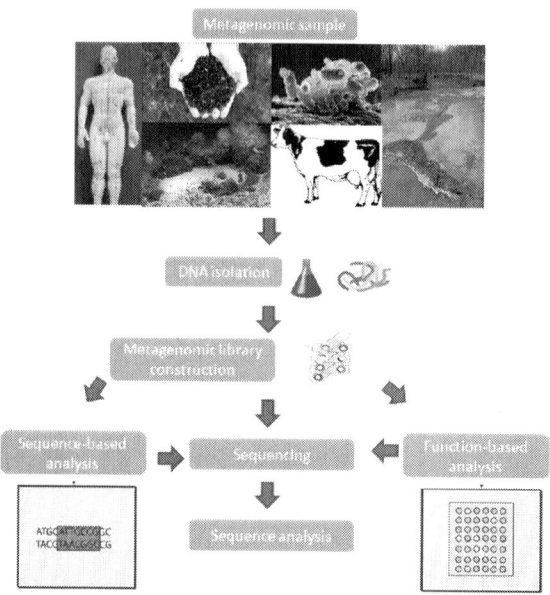

Figure 2: Schematic representation of the different steps for metagenomic analysis.

One of the first studies using metagenomics to study microbial degradation of aromatic compounds was performed by Suenaga and colleagues [100], who constructed a metagenomic library from activated sludge for industrial wastewater. The library was functionally screened for extradiol dioxygenase activities (enzymes for aromatic degradation) and 38 clones were subjected to sequencing analysis [101]. As a result, various types of gene subsets were identified that were not similar to the previously reported pathways performing complete degradation. Moreover, the authors discussed the fact that aromatic compounds in the environment may be degraded through the concerted action of various fragmented pathways. Sierra-Garcia [101] reported the organization of hydrocarbon degradation genes of selected metagenomic fosmid clones derived from a metagenomic library from Brazilian petroleum reservoir and functional screening for hydrocarbon degradation activities. The author found many putative proteins of different aerobic and anaerobic well described catabolic pathways, however the complete catabolic pathways described for hydrocarbon degradation in previous studies were absent in the fosmid clones. Instead, the metagenomic fragments comprised genes belonging to different pathways, showing novel gene arrangements where hydrocarbon compounds were degraded through the concerted actions of these fragmented pathways. These results suggest that there are marked differences between the degradation genes found in microbial communities derived from enrichments of oil reservoir sample and those that have been previously identified in bacteria isolated from contaminated or pristine environments.

However, function-based screening of metagenomic libraries for xenobiotic degradation genes is often considered problematic because of insufficient and biased expression of the heterologous genes in the host *Escherichia coli* [99]. Only a few efforts have been made to solve these problems. In Uchiyama et al. [103], a novel method for function-driven screening is described, which was termed substrate-induced gene expression screening (SIGEX). This high-throughput screening approach employs an operon trap gfp expression vector in combination with fluorescence-activated cell sorting. The screening is based on the fact that catabolic-gene expression is induced mainly by specific substrates and is often controlled by regulatory elements located close to catabolic genes [103]. Using this approach, Uchiyama et al. [103] isolated aromatic-hydrocarbon-induced genes from a metagenomic library derived from groundwater. In Ono et al. [104] another screening strategy was based on functional complementation of a *Pseudomonas putida* host strain containing a naphthalene degrading pathway devoid of the naphthalene dioxygenase (NDO) encoding gene. Two clones were able to restore the ability of the host strain to use naphthalene as a sole carbon source and their genes were similar but no identical to already known operons. The authors refer to the use of other host strains for the construction of metagenomic libraries instead of the well-established *E. coli* as a simpler and economical way to perform function-driven screening in comparison to other reported systems such as SIGEX [103].

In the context of this chapter, several aspects of the hydrocarbon degradation need to be studied to obtain a comprehensive overview of the biodegradation processes that take place in oil reservoirs or petroleum impacted environments. These studies should take into consideration the high diversity of enzymes capable of initiating such metabolism as well as the implementation of integrated studies combining culture and molecular techniques, linking with metabolomics or compound-specific isotope analysis and microcosm studies for a better resolution of in situ microbial activity in petroleum reservoirs.

CONCLUSIONS AND RESEARCH NEEDS

The understanding about biodegraded petroleum reservoirs have advanced considerably in recent years, but the organisms responsible for the *in situ* activity and a quantitative understanding of the factors which control in-reservoir oil biodegradation remain far from complete. The inaccessibility of petroleum reservoirs and inherent difficulties of microbiological sampling from commercially operating oil wells have required a multidisciplinary approach to delineating the study of subsurface petroleum biodegradation, and to date there are still prevailing paradigms relating to hydrocarbon biodegradation processes. This multidisciplinary approach to study *in situ* petroleum degradation should consider molecular biology, microbiology, and geological and geochemical parameters in order to establish the key organisms, biochemical reactions and mechanisms involved in such complex associations. Indeed, the isolation of anaerobic microorganisms capable of utilizing hydrocarbons is essential for a comprehensive understanding of their role and behavior in anoxic habitats and their complex interactions within methanogenic hydrocarbon-degrading communities. In addition, novel approaches, combining functional metagenomics, transcriptomics, metabolomics and other molecular surveys in microcosms are urgently required to better allow access to a more realistic phylogenetic and metabolic diversity governing oil biodegradation in petroleum reservoirs.

REFERENCES

1. W Roling, 2003The microbiology of hydrocarbon degradation in subsurface petroleum reservoirs: perspectives and prospects. *Res Microbiol.* 154(5), 321-328.
2. I. M Head, D. M Jones, S. R Larter, 2003Biological activity in the deep subsurface and the origin of heavy oil. *Nature,* 426(6964), 344-52.
3. W Ismail, and J Gescher, 2012Epoxy coenzyme a thioester pathways for degradation of aromatic compounds. *Appl Environ Microbiol.* 7815504351
4. D. P Chandler, S. M Li, C. M Spadoni, G. R Drake, D. L Balkwill, J. K Fredrickson, F. J Brockman, 1997A molecular comparison of culturable aerobic heterotrophic bacteria and 16S rDNA clones derived from a deep subsurface sediment. *FEMS Microbiol Ecol.* 23131144
5. M. B Leigh, V. H Pellizari, O Uhlik, R Sutka, J Rodrigues, N. E Ostrom, et al2007Biphenyl-utilizing bacteria and their functional genes in a pine root zone contaminated with polychlorinated biphenyls (PCBs). *ISME J* 1134148
6. V De Lorenzo, 2008Systems biology approaches to bioremediation. *Curr Opin Biotechnol* 19579589
7. C Jeon, W Park, P Padmanabhan, C Derito, J Snape, E Madsen, 2003Discovery of a bacterium, with distinctive dioxygenase, that is responsible for*in situ* biodegradation in contaminated sediment. *Proc Natl Acad Sci USA 100*1359113596
8. R Witzig, H Junca, H. J Hecht, D. H Pieper, 2006Assessment of toluene/biphenyl dioxygenase gene diversity in benzene-polluted soils: links between benzene biodegradation and genes similar to those encoding isopropylbenzene dioxygenases. *Appl Environ Microbiol 72*35043514

9. J Foght, 2010Microbial comminities in oil shales, biodegraded and heavy oil reservoirs, and bitumen deposits. In: K. N. Timmis (Ed.) *Handbook of Hydrocarbon and Lipid Microbiology*. Berlin, Heidelberg: Springer Berlin Heidelberg.

10. N. K Birkeland, 2004The microbial diversity of deep subsurface oil reservoirs. *Stud Surface Sci Catal* 151385403

11. I. M Head, C. M Aitken, N. D Gray, A Sherry, J. J Adams, D. M Jones, A. K Rowan, et al2010Hydrocarbon degradation in petroleum reservoirs. In: K. N. Timmis (Ed.) *Handbook of Hydrocarbon and Lipid Microbiology*. Berlin, Heidelberg: Springer Berlin Heidelberg.

12. L Haridon, S Reysenbach, A. L Glenat, P Prieur, D Jeanthon, C. (1995Hot subterranean biosphere in a continental oil reservoir. *Nature* 377223224

13. G. S Grassia, K. M Mclean, P Glenat, J Bauld, A. J Sheehy, 1996A systematic survey for thermophilic fermentative bacteria and archaea in high temperature petroleum reservoirs. *FEMS Microbiol* Ecol 214758

14. M Magot, B Ollivier, B. K. C Patel, 2000Microbiology of petroleum reservoirs. *Antonie van Leeuwenhoek*. 772103116

15. V. J Orphan, L. T Taylor, D Hafenbradl, and E. F Delong, 2000Culture-dependent and culture-independent characterization of microbial assemblages associated with high-temperature petroleum reservoirs. *Appl Environ Microbiol*. 66270011

16. C. M Aitken, D. M Jones, S. R Larter, 2004Anaerobic hydrocarbon biodegradation in deep subsurface oil reservoirs. *Nature*, 43170062914

17. N. D Gray, A Sherry, C Hubert, J Dolfing, I. M Head, 2010Methanogenic degradation of petroleum hydrocarbons in subsurface environments remediation, heavy oil formation, and energy recovery. *Adv Appl Microbiol*. 7213761

18. F Widdel, R Rabus, 2001Anaerobic biodegradation of saturated and aromatic hydrocarbons. *Curr Opin Biotechnol* 12259276

19. L. M Gieg, K. E Duncan, J. M Suflita, 2008Bioenergy production via microbial conversion of residual oil to natural gas. Appl Environ Microbiol 7430223029

20. D Jones, I Head, N Gray, J Adams, A Rowan, C Aitken, B Bennett, et al2007Crude-oil biodegradation via methanogenesis in subsurface petroleum reservoirs. *Nature*, 451(7175), 176-180.
21. K Zengler, H. H Richnow, R Rossello-mora, W Michaelis, F Widdel, 1999Methane formation from long chain alkanes by anaerobic microorganisms.*Nature* 401266269
22. J Beeder, T Torsvik, and T Lien, 1995*Thermodesulforhabdus norvegicus* gen. nov., sp. nov., a novel thermophilic sulfate-reducing bacterium from oil field water. *Arch. Microbiol* 164331336
23. G. N Rees, G. S Grassia, A. J Sheehy, P. P Dwivedi, B. K. C Patel, 1995*Desulfacinum infernum* gen. nov., sp. nov., a thermophilic sulfate-reducing bacterium from a petroleum reservoir. *Int. J. Syst. Bacteriol* 458589
24. [24] A Grabowski, O Nercessian, F Fayolle, D Blanchet, C Jeanthon, 2005Microbial diversity in production waters of a low-temperature biodegraded oil reservoir. *FEMS microbiology ecology*, 54(3), 427-43.
25. M Magot, G Ravot, X Campaignolle, B Ollivier, B. K Patel, M. L Fardeau, P Thomas, J. L Crolet, J. L Garcia, 1997Dethiosulfovibrio peptidovorans gen. nov., sp. nov., a new anaerobic, slightly halophilic, thiosulfate-reducing bacterium from corroding offshore oil wells. *Int. J. Syst. Bacteriol.* 47818824
26. R. K Nilsen, T Torsvik, T Lien, 1996Desulfotomaculum thermocisternum sp. nov., a sulfate reducer isolated from a hot North Sea oil reservoir. *Int. J. Syst. Bacteriol.* 46397402
27. G Ravot, M Magot, B Ollivier, B. K. C Patel, E Ageron, P. A. D Grimont, P Thomas, J. L Garcia, 1997*Haloanaerobium congolense* sp. nov., an anaerobic, moderately halophilic, thiosulfate- and sulfur-reducing bacterium from an African oil field. *FEMS Microbiol. Lett.* 1478188
28. E Miranda-tello, M. L Fardeau, L Fernandez, F Ramirez, J. L Cayol, P Thomas, J. L Garcia, B Ollivier, 2003*Desulfovibrio capillatus* sp. nov., a novel sulfatereducing bacterium isolated from an oil field separator located in the Gulf of Mexico. *Anaerobe* 997103
29. T. N Nazina, T. P Tourova, A. B Poltaraus, E. V Novikova, A. A Grigoryan, A. E Ivanova, et al2001Taxonomic study of aerobic

thermophilic bacilli: Descriptions of *Geobacillus subterraneus* gen. nov., sp. nov. and *Geobacillus uzenensis* sp. nov. from petroleum reservoirs and transfer of *Bacillus stearothermophilus, Bacillus hermocatenulatus, Bacillus thermoleovorans, Bacillus kaustophilus, Bacillus thermoglucosidasius* and *Bacillus thermodenitrificans* to *Geobacillus* as the new combinations *G. stearothermophilus, G. thermocatenulatus, G. thermoleovorans, G. kaustophilus, G. thermoglucosidasius* and *G. thermodenitrificans*. *Int. J. Syst. Evol. Microbiol.* 51433446

30. M. L Miroshnichenko, H Hippe, E Stackebrandt, N. A Kostrikina, N. A Chernyh, C Jeanthon, T. N Nazina, S. S Belyaev, E. A Bonch-osmolovskaya, 2001Isolation and characterization of *Thermococcus sibiricus* sp. nov. from a Western Siberia high-temperature oil reservoir. *Extremophiles*.58591

31. G Ravot, M Magot, M. L Fardeau, B. K. C Patel, P Thomas, J. L Garcia, B Ollivier, 1999*Fusibacter paucivorans* gen. nov., sp. nov., an anaerobic, thiosulfate-reducing bacterium from an oil-producing well. *Int. J. Syst. Bacteriol.* 4911411147

32. H Dahle, and N. K Birkeland, 2006*Thermovirga lienii* gen. nov., sp. nov., a novel moderately thermophilic, anaerobic, amino-acid-degrading bacterium isolated from a North Sea oil well. *Int. J. Syst. Evol. Microbiol.* 5615391545

33. R. K Nilsen, and T Torsvik, 1996*Methanococcus thermolithotrophicus* isolated from North sea oil field reservoir water. *Appl. Environ. Microbiol.* 62728731

34. M Magot, 2005Indigenous microbial communities in oil fields. In B. Ollivier and M. Magot, (Eds.) Petroleum microbiology. 2134ASM, Washington, DC.

35. N. L Söhngen, 1913Benzin, Petroleum, Paraffinöl und Paraffin als Kohlenstoff- und Energiequelle für Mikroben. *Zentr Bacteriol Parasitenk Abt II* 37595609

36. F Widdel, and F Musat, 2010Diversity and common principles in enzymatic activation of hydrocarbons. In: K. N. Timmis (Ed.) *Handbook of Hydrocarbon and Lipid Microbiology*. Berlin, Heidelberg: Springer Berlin Heidelberg.

37. M Boll, and J Heider, 2010Anaerobic Degradation of Hydrocarbons: Mechanisms of C-H-Bond activation in the absence of oxygen.

In: K. N. Timmis (Ed.) *Handbook of Hydrocarbon and Lipid Microbiology*. Berlin, Heidelberg: Springer Berlin Heidelberg.

38. J Foght, 2008Anaerobic biodegradation of aromatic hydrocarbons: pathways and prospects. *J Mol Microbiol Biotechnol*. 15(2-3): 93-120.
39. J. B Van Beilen, and E. G Funhoff, 2007Alkane hydroxylases involved in microbial alkane degradation. *Appl Microbiol Biotechnol*. 7411321
40. A Wentzel, T. E Ellingsen, H. K Kotlar, S. B Zotchev, M Throne-holst, 2007Bacterial metabolism of long-chain n-alkanes. *Appl Microbiol Biotechnol*. 76612091221
41. F Rojo, 2009*Degradation of alkanes by bacteria. Environmental microbiology*.
42. J. D Van Hamme, A Singh, O. P Ward, 2003Recent advances in petroleum microbiology. *Microbiol Mol Biol Rev*. 674503549
43. F Rojo, 2010Enzymes for Aerobic Degradation of Alkanes. In K. N. Timmis (Ed.), *Handbook of Hydrocarbon and Lipid Microbiology* (781Berlin, Heidelberg: Springer Berlin Heidelberg.
44. R Margesin, D Labbe, F Schinner, C Greer, L Whyte, 2003Characterization of hydrocarbon-degrading microbial populations in contaminated and pristine alpine soils. *Appl Environ Microbiol*. 69630853092
45. E Kuhn, G. S Bellicanta, V. H Pellizari, 2009New alk genes detected in Antarctic marine sediments. *Environ Microbiol*. 113669673
46. J. M Salminen, P. M Tuomi, K. S Jorgensen, 2008Functional gene abundances (*nahAc, alkB, xylE*) in the assessment of the efficacy of bioremediation. *Appl Biochem Biotechnol* 151638652
47. N Hamamura, M Fukui, D. M Ward, W. P Inskeep, 2008Assessing soil microbial populations responding to crude-oil amendment at different temperatures using phylogenetic, functional gene (*alkB*) and physiological analyses. *Environ Sci Technol* 4275807586
48. J. B Van Beilen, M. G Wubbolts, B Witholt, 1994Genetics of alkane oxidation by *Pseudomonas oleovorans*. *Biodegradation* 561174

49. R Marchant, F. H Sharkey, I. M Banat, T. J Rahman, A Perfumo, 2006The degradation of n-hexadecane in soil by thermophilic geobacilli. *FEMS Microbiol Ecol.* 561444
50. J. B Van Beilen, Z Li, W. A Duetz, T. H. M Smits, B Witholt, 2003Diversity of Alkane Hydroxylase Systems in the Environment. *Oil Gas Sci Technol.*584427440
51. T. P Tourova, T. N Nazina, E. M Mikhailova, T. A Rodionova, A. N Ekimov, A. V Mashukova, A. B Poltaraus, 2008alkB homologs in thermophilic bacteria of the genus *Geobacillus*. *Mol Biol.* 422217226
52. W Li, L. Y Wang, R. Y Duan, J. F Liu, J. D Gu, B. Z Mu, 2012Microbial community characteristics of petroleum reservoir production water amended with n-alkanes and incubated under nitrate-, sulfate-reducing and methanogenic conditions. *Inter Biodeterior Biodegradation.* 698796
53. R Vilchez-vargas, H Junca, D. H Pieper, 2010Metabolic networks, microbial ecology and "omics" technologies: towards understanding in situ biodegradation processes. *Environ Microbiol.* 1230893104
54. L Feng, W Wang, J Cheng, Y Ren, G Zhao, C Gao, Y Tang, et al2007Genome and proteome of long-chain alkane degrading Geobacillus thermodenitrificans NG80-2 isolated from a deep-subsurface oil reservoir. *Proc Natl Acad Sci U S A.* 1041356027
55. M Carmona, M Zamarro, B Blazquez, G Durante-rodriguez, J Juarez, J Valderrama, *et al*2009Anaerobic catabolism of aromatic compounds: a genetic and genomic view. *Microbiol Mol Biol Rev.* 7371133
56. M. V Brennerova, J Josefiova, V Brenner, D. H Pieper, H Junca, 2009Metagenomics reveals diversity and abundance of meta-cleavage pathways in microbial communities from soil highly contaminated with jet fuel under air-sparging bioremediation. *Environ Microbiol.* 119221627
57. D Pérez-pantoja, B González, D. H Pieper, 2010Aerobic degradation of aromatic hydrocarbons. In: K. N. Timmis (Ed.) *Handbook of Hydrocarbon and Lipid Microbiology*. Berlin, Heidelberg: Springer Berlin Heidelberg.

58. Y Jouanneau, 2010Oxidative inactivation of ring cleavage extradiol dioxigenases: mechanism and ferredoxin mediated reactivation. In: K. N. Timmis (Ed.) *Handbook of Hydrocarbon and Lipid Microbiology*. Berlin, Heidelberg: Springer Berlin Heidelberg.
59. H Junca, and D. H Pieper, 2003Functional gene diversity analysis in BTEX contaminated soils by means of PCR-SSCP DNA fingerprinting: comparative diversity assessment against bacterial isolates and PCR-DNA clone libraries. *Environ Microbiol.* 6295110
60. F Mehboob, H Junca, G Schraa, A. J. M Stams, 2009Growth of Pseudomonas chloritidismutans AW-1(T) on n-alkanes with chlorate as electron acceptor. *Appl Microbiol Biotechnol.* 83473947
61. G Fuchs, 2008Anaerobic metabolism of aromatic compounds. *Ann N Y Acad Sci.* 11258299
62. M Kube, J Heider, J Amann, P Hufnagel, S Kühner, A Beck, R Reinhardt, et al2004Genes involved in the anaerobic degradation of toluene in a denitrifying bacterium, strain EbN1. *Arch Microbiol.* 181318294
63. M Boll, G Fuchs, J Heider, 2002Anaerobic oxidation of aromatic compounds and hydrocarbons. *Curr Opin Chem Biol.* 6560411
64. f. M Kaser, and J. D Coates, 2010Nitrate, Perchlorate and Metal respirers. In: K. N. Timmis (Ed.) *Handbook of Hydrocarbon and Lipid Microbiology*. Berlin, Heidelberg: Springer Berlin Heidelberg.
65. O Grundmann, A Behrends, R Rabus, J Amann, T Halder, J Heider, F Widdel, 2008Genes encoding the candidate enzyme for anaerobic activation of n-alkanes in the denitrifying bacterium, strain HxN1. *Environ Microbiol.* 10237685
66. S. R Kane, H. R Beller, T. C Legler, R. T Anderson, 2002Biochemical and genetic evidence of benzylsuccinate synthase in toluene-degrading, ferric iron-reducing *Geobacter metallireducens*. *Biodegradation*, 13214954
67. C Winderl, S Schaefer, T Lueders, 2007Detection of anaerobic toluene and hydrocarbon degraders in contaminated aquifers using benzylsuccinate synthase (bssA) genes as a functional marker. *Environ Microb*iol 910351046

68. C Winderl, B Anneser, C Griebler, R. U Meckenstock, T Lueders, 2008Depth resolved quantification of anaerobic toluene degraders and aquifer microbial community patterns in distinct redox zones of a tar oil contaminant plume. *Appl Environ Microbiol* 74792801
69. M Staats, M Braster, W. F. M Roling, 2011Molecular diversity and distribution of aromatic hydrocarbon-degrading anaerobes across a landfill leachate plume. *Environ Microbiol* 1312161227
70. B Wawrik, M Mendivelso, V. A Parisi, J. M Suflita, I. A Davidova, C. R Marks, J. D Van Nostrand, Y Liang, J Zhou, B. J Huizinga, et al2012Field and laboratory studies on the bioconversion of coal to methane in the San Juan Basin. *FEMS Microbiol Ecol.* 812642
71. A. V Callaghan, I. A Davidova, K Savage-ashlock, V. A Parisi, L. M Gieg, J. M Suflita, J. J Kukor, et al2010Diversity of benzyl- and alkylsuccinate synthase genes in hydrocarbon-impacted environments and enrichment cultures. *Environ Sci Technol.* 4419728794
72. F Widdel, and O Grundmann, 2010Biochemistry of the anaerobic degradation of non-methane alkanes. In: K. N. Timmis (Ed.) *Handbook of Hydrocarbon and Lipid Microbiology*. Berlin, Heidelberg: Springer Berlin Heidelberg.
73. V Grossi, C Cravolaureau, R Guyoneaud, A Ranchoupeyruse, A Hirschlerrea, 2008Metabolism of n-alkanes and n-alkenes by anaerobic bacteria: A summary. *Org Geochem.* 39811971203
74. L. M Gieg, I. A Davidova, K. E Duncan, J. M Suflita, 2010Methanogenesis, sulfate reduction and crude oil biodegradation in hot Alaskan oilfields.*Environ Microbiol.* 1211307486
75. S. M Mbadinga, K. P Li, L Zhou, L. Y Wang, S Yang, Z Liu, J. F Gu, J.D., et al2012Analysis of alkane-dependent methanogenic community derived from production water of a high-temperature petroleum reservoir. *Appl Microbiol Biotechnol.* 96253142
76. L. M Gieg, and J. M Suflita, 2002Detection of anaerobic metabolites of saturated and aromatic hydrocarbons in petroleum-contaminated aquifers.*Environ. Sci. Technol.* 361737553762
77. K. E Duncan, L. M Gieg, V. A Parisi, R. S Tanner, J. M Suflita, Green Tringe, S., Bristow, J. (2009Biocorrosive thermophilic microbial

communities in Alaskan North Slope oil facilities. *Environ Sci Technol* 4379777984

78. A. V Callaghan, B Wawrik, NlChadhain, S.M., Young, L.Y., Zylstra, G.J. (2008Anaerobic alkane-degrading strain AK-01 contains two alkylsuccinate synthase genes. *Biochem Biophys Res Commun.* 366142148

79. J Zedelius, R Rabus, O Grundmann, I Werner, D Brodkorb, F Schreiber, P Ehrenreich, A Behrends, H Wilkes, M Kube, R Reinhardt, F Widdel, 2010Alkane degradation under anoxic conditions by a nitrate-reducing bacterium with possible involvement of the electron acceptor in substrate activation. *Environ Microbiol Rep.* 31125135

80. S. M Mbadinga, L. Y Wang, L Zhou, J. F Liu, J. D Gu, B. Z Mu, 2011Microbial communities involved in anaerobic degradation of alkanes. *Inter Biodeterior Biodegradation.* 651113

81. C So, C Phelps, L Young, 2003Anaerobic transformation of alkanes to fatty acids by a sulfate-reducing bacterium, strain Hxd3. *Appl Environ.* 69738923900

82. A. V Callaghan, M Tierney, C. D Phelps, L. Y Young, 2009Anaerobic biodegradation of n-hexadecane by a nitrate-reducing consortium. *Appl Environ Microbiol* 7513391344

83. Head, I., Gray, N., Aitken, C., Sherry, A., Jones, M., Larter, S. (2010). Hydrocarbon activation under sulfate-reducing and methanogenic conditions proceeds by different mechanisms. Geophysical Research Abstracts 12 (EGU General Assembly 2010

84. V Torsvik, J Goksoyr, F. L Daae, 1990High diversity in DNA of soil bacteria. *Appl Environ Microbiol* 56782787

85. R. I Amann, W Ludwig, K. H Schleifer, 1995Phylogenetic identification and in situ detection of individual microbial cells without cultivation.*Microbiol Rev* 59143169

86. V Torsvik, F. L Daae, R. A Sandaa, L Øvreås, 1998Novel techniques for analyzing microbial diversity in natural and perturbed environments. *J Biotechnol* 645362

87. E Kellenberger, 2001Exploring the unknown: the silent revolution of microbiology. *EMBO reports*, 2(1), 2-5.

88. V. J Orphan, S. K Goffredi, E. F Delong, J. R Boles, 2003Geochemical influence on diversity and microbial processes in high temperature oil reservoirs. *Geomicrobiol J* 20295311
89. T. N Nazina, N. M Shestakova, Grigor'yan, A.A., Mikhailova, E.M., Tourova, T.P., Poltaraus, A.B., et al. (2006Phylogenetic diversity and activity of anaerobic microorganisms of high-temperature horizons of the Dagang oil field (P.R. China). *Microbiology* 755565
90. D Mayumi, H Mochimaru, H Yoshioka, S Sakata, H Maeda, Y Miyagawa, M Ikarashi, et al2011Evidence for syntrophic acetate oxidation coupled to hydrogenotrophic methanogenesis in the high-temperature petroleum reservoir of Yabase oil field (Japan). *Environ Microbiol.* 13819952006
91. J. L Garcia, B. K Patel, B Ollivier, 2000Taxonomic, phylogenetic, and ecological diversity of methanogenic Archaea. *Anaerobe* 6205226
92. S. H Zinder, and M Koch, 1984Non-acetoclastic methanogenesis from acetate: acetate oxidation by a thermophilic syntrophic coculture. *Arch Microbiol 138*263272
93. A Schnurer, F. P Houwen, B. H Svensson, 1994Mesophilic syntrophic acetate oxidation during methane formation by a triculture at high ammonium concentration. *Arch Microbiol 162*7074
94. S Hattori, Y Kamagata, S Hanada, H Shoun, 2000*Thermacetogenium phaeum* gen. nov., sp. nov., a strictly anaerobic, thermophilic, syntrophic acetate-oxidizing bacterium. *Int J Syst Evol Microbiol 50*16011609
95. M Balk, J Weijma, A. J Stams, 2002*Thermotoga lettingae* sp. nov., a novel thermophilic, methanoldegrading bacterium isolated from a thermophilic anaerobic reactor. *Int J Syst Evol Microbiol 52*13611368
96. T Barth, 1991Organic-acids and inorganic-ions in waters from petroleum reservoirs, Norwegian continental-shelf: a multivariate statistical-analysis and comparison with American reservoir formation waters. *Appl Geochem* 6115
97. T. R Silva, L. C. L Verde, Santos Neto, E.V., Oliveira, V.M. (2012Diversity analyses of microbial communities in petroleum

samples from Brazilian oil fields. Inter Biodeterior Biodegradation doi:10.1016/j.ibiod.2012.05.005.

98. J Dolfing, S. R Larter, I. M Head, 2008Thermodynamic constraints on methanogenic crude oil biodegradation. *ISME J* 2442452

99. T Uchiyama, and K Miyazaki, 2009Functional metagenomics for enzyme discovery: challenges to efficient screening. *Curr Opin Biotechnol*.206616622

100. H Suenaga, T Ohnuki, K Miyazaki, 2007Functional screening of a metagenomic library for genes involved in microbial degradation of aromatic compounds. *Environ Microbiol*. 9922892297

101. H Suenaga, Y Koyama, M Miyakoshi, R Miyazaki, H Yano, M Sota, Y Ohtsubo, et al2009Novel organization of aromatic degradation pathway genes in a microbial community as revealed by metagenomic analysis. *ISME J*. 312133548

102. I. N Sierra-garcia, Caracterização estrutural e funcional de genes de degradação de hidrocarbonetos originados de metagenoma microbiano de reservatório de petróleo. M SC. Thesis. Universidade Estadual de Campinas; 2011

103. T Uchiyama, T Abe, T Ikemura, K Watanabe, 2005Substrate-induced gene-expression screening of environmental metagenome libraries for isolation of catabolic genes. *Nat Biotechnol*. 2318893

104. A Ono, R Miyazaki, M Sota, Y Ohtsubo, Y Nagata, M Tsuda, 2007Isolation and characterization of naphthalene-catabolic genes and plasmids from oil-contaminated soil by using two cultivation-independent approaches. *Appl Microbiol Biotechnol*. 74250110

CITATION

Isabel Natalia Sierra-Garcia and Valéria Maia de Oliveira (2013). Microbial Hydrocarbon Degradation: Efforts to Understand Biodegradation in Petroleum Reservoirs, Biodegradation - Engineering and Technology, Dr. Rolando Chamy (Ed.), ISBN: 978-953-51-1153-5, InTech, DOI: 10.5772/55920.

Citations

CHAPTER 1

Snehanshu Pal and T. K. Kundu, "Theoretical Study of Hydrogen Bond Formation in Trimethylene Glycol-Water Complex," ISRN Physical Chemistry, vol. 2012, Article ID 570394, 12 pages, 2012, doi:10.5402/2012/570394.

CHAPTER 2

Kaufui V. Wong and Omar De Leon, "Applications of Nanofluids: Current and Future," Advances in Mechanical Engineering, vol. 2010, Article ID 519659, 11 pages, 2010. doi:10.1155/2010/519659.

CHAPTER 3

J. K. Nørskov, T. Bligaard, J. Rossmeisl, and C. H. Christensen, Towards the Computational Design of Solid Catalysts, doi:10.1038/nchem.121.

CHAPTER 4

Zhou Yazhou and Yin Daiyin, Study and Application of Numerical Simulation of Deep Profile Control with Weak Gel, http://dx.doi.org/10.14257/ijca.2013.6.5.26.

CHAPTER 5

Dali Hou, Pingya Luo, Lei Sun, Yong Tang, and Yi Pan, "Study on Nonequilibrium Effect of Condensate Gas Reservoir with Gaseous Water under HT and HP Condition," Journal of Chemistry, vol. 2014, Article ID 295149, 8 pages, 2014. doi:10.1155/2014/295149.

CHAPTER 6

Jian-Yi Liu, Jing Zhang, Yan-Li Liu, Xiao-Hua Tan, and Jie Zhang, "Experimental and Modeling Studies on the Prediction of Gas Hydrate Formation," Journal of Chemistry, Article ID 198176, in press.

CHAPTER 7

G. Steyl and G. J. van Tonder (2013). Hydrochemical and Hydrogeological Impact of Hydraulic Fracturing in the Karoo, South Africa,

Effective and Sustainable Hydraulic Fracturing, Dr. Rob Jeffrey (Ed.), ISBN: 978-953-51-1137-5, InTech, DOI: 10.5772/56310.

CHAPTER 8

Lijun You, Kunlin Xue, Yili Kang, Yi Liao, and Lie Kong, "Pore Structure and Limit Pressure of Gas Slippage Effect in Tight Sandstone," The Scientific World Journal, vol. 2013, Article ID 572140, 7 pages, 2013, doi:10.1155/2013/572140.

CHAPTER 9

Ganapati D. Yadav and Jyoti B. Sontakke (2013). Methods for Separation, Recycling and Reuse of Biodegradation Products, Biodegradation - Engineering and Technology, Dr. Rolando Chamy (Ed.), ISBN: 978-953-51-1153-5, InTech, DOI: 10.5772/56241.

Index

A

Acrylonitrile-butadiene styrene (ABS) 198
Amorphous organic matter (AOM) 3
Argonne National Laboratory (ANL) 198

B

Biodegradation 117, 118, 125, 136, 141, 142, 144, 145
body-of-proof (BP) 49
Body-of-proof (BP) 49, 50

C

Carbon source 178
Carbon steel 42
Cellular respiration process (CSP) 177
China Automotive Technology and Research Center (CAT-ARC) 59
Chlorinated hydrocarbon (CHC) 182
Compressed natural gas (CNG 56
Crystal nucleus growth 102, 103

D

Drainage gas recovery 23

E

Economical proces 179

Expanded granular sludge bed (EGSB) 181

G

Gaseous water 77, 78, 79, 82, 84, 85, 86, 87, 88, 93, 96
Gas slip effect 149, 150, 152
Gas-to-liquid- (GTL-) 56
Genetically engineered microorganisms (GEMs) 133
Genetically modified (GM) 135
Geochemical data 2
Good economic 24
Greenhouse gas (GHG) 55, 59

H

High-density polyethylene (HDPE) 197
High-impact polystyrene (HIPS) 198
High-performance ion-exchange chromatography (HPIC) 84
Hydrate formation 102, 103, 104, 105, 107, 108, 109, 110, 112, 113, 114
Hydroquinone (HQ) 185

I

Industrial painting 43

L

Large number 178, 182
Life-cycle analyses (LCAs) 59
Life-cycle emission model (LEM) 59
Liquefied natural gas (LNG) 56

M

Marine Corps Air Ground Combat Center (MCAGCC) 130
Methane monooxygenase (MMO) 183
Microorganism 171, 172, 173, 176, 177, 178, 183, 187, 189, 190, 195, 201, 206, 213

N

Natural gas 24, 26, 27, 102, 103, 108, 113
Nutrient 121

O

Oil density of synthetic 84
Organic fraction of municipal solid waste (OFMSW) 181
Organic matter (OM) 18

P

Pentachlorophenol (PCP) 186
Perchloroethylene (PCE) 182
Petrographic analyse 3
Petroleum-based product 116
Petroleum hydrocarbons (PHCs) 182
Physical value 35
Plasticizer 188, 212
P-Nitrophenol (PNP) 185
Polychlorinated biphenyls (PCBs) 176
Polyethylene terephthalate (PET) 194, 199
Polyhydroxyalkanoates (PHAs) 190
Polylactic acid (PLA) 195
Polyvinyl alcohol (PVA) 190
Polyvinyl chloride (PVC) 200
Precipitate lag 79

Index

Production Index (PI) 8
Pump to wheels (PTW) 64

R

Recovery Plastics International (RPI) 197, 198

S

Software applicant 36
Solids retention time (SRT) 180, 181

T

Tetrachloroethylene (TCE) 183
Total organic carbon (TOC) 7
Total petroleum hydrocarbon (TPH) 132
Trichloroethane (TCA) 182
Trichloroethylene (TCE) 182

Tsinghua life-cycle analysis model (TLCAM) 55, 60

U

Upflow anaerobic sludge blanket (UASB) 181

V

Vinyl chloride (VC) 183
Volatile organic compounds (VOCs) 170
Volatile solids reduction (VSR) 180

W

Wastewater 179, 180, 181, 184, 192, 205, 212
Well to pump (WTP) 64
Well-to-wheels (WTW) 56, 60